Civil Engineering Heritage Wales and West Central England

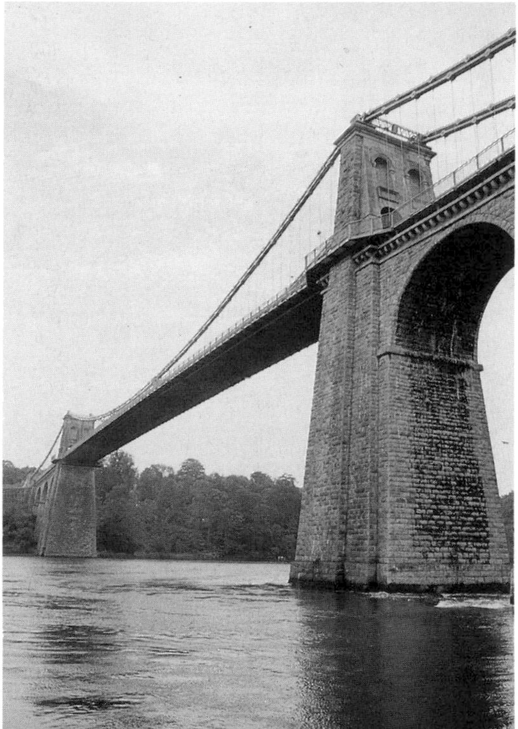

Edited by Roger Cragg, BSc, MSc, CEng, MICE, MIHT

Other books in the Civil Engineering Heritage series:
Northern England. R. W. Rennison
Southern England. R. A. Otter
Eastern and Central England. E. A. Labrum

Future titles in the series:
Scotland
Ireland
London

Wales and West Central England
Published for the Institution of Civil Engineers by
Thomas Telford Publishing, Thomas Telford Services Ltd, 1 Heron Quay, London E14 4JD

First published 1986
Second edition 1997

A CIP record exists for this book

ISBN 0 7277 2576 9

Typeset in Palatino 10/11.5 pt using Ventura by Katy Carter
Printed by Interprint Limited, Malta

Preface

In 1971 the Institution of Civil Engineers set up the Panel for Historical Engineering Works, one of its objectives being the creation and maintenance of an archive containing records of civil engineering structures of historical interest and significance. A further objective of the Panel is to enhance awareness of our engineering heritage and to persuade others to take an interest in civil engineering works which illustrate the history and development of the profession.

The series *Civil Engineering Heritage* seeks to satisfy the latter objective by drawing on the Panel's archive and research and by making this information available to a wider public.

In 1981 the first volume in the series, *Northern England*, was published, followed in 1986 by the first edition of this book, then entitled *Wales and Western England* and edited by W. J. Sivewright. Subsequently two other volumes were published in 1994, covering *Southern England* and *Eastern and Central England*, followed by an extensively revised and enlarged second edition of *Northern England* in 1996.

In preparing this revised edition of *Wales and West Central England* the opportunity has been taken to include a number of new items and to delete a small number of structures which have been demolished in the years since the publication of the first edition. In addition, the items brought forward from the first edition have been updated where necessary in the light of recent developments.

Britain has, for the past 250 years, been at the forefront of engineering design and innovation. Within the pages of this volume will be found many examples of the skills of the civil engineer, both past and present. They illustrate the wide ranging nature of the profession in providing the infrastructure for contemporary society, covering transport systems, from canals and railways to modern motorways and airports; public health, from water supply to sewage treatment; power supply, from watermills to electricity generating stations; and many other aspects.

As Editor, I must gratefully acknowledge the assistance in the preparation of this volume of many people, not least the late W. J. Sivewright whose work as Editor of the first edition provided a firm foundation upon which to build. The Panel Members who were the principal contributors to the first edition of this book were Paul Dunkerley, Owen Gibbs, the late Roy Hughes, the late W. J. Sivewright and myself. For the second edition, contributing Panel members were Paul Dunkerley, David Greenfield and Gareth George, together with Owen Gibbs. Valuable contributions were also made by Brian Haskins, Roy P. Williams and many others whose assistance is gratefully acknowledged. I am also grateful for the assistance of many local authorities, public utilities and companies in supplying information and comment.

My particular thanks are due to Bryan O'Loughlin, Vice-Chairman of the Panel and to Michael Chrimes, Librarian of the Institution, for their invaluable help in checking the text and in offering much helpful advice drawn from their expertise and knowledge of the history of civil engineering. The assistance of Mary Murphy, Archivist of the Institution and Carol Arrowsmith in patiently providing information is also gratefully acknowledged.

Finally, acknowledgement is made to the Ordnance Survey for permitting the reproduction of maps and the use of national grid references.

Roger Cragg

Contents

Introduction	1
Metric equivalents	3
1. North Wales	4
2. Mid-Wales	52
3. South Wales	68
4. The Bristol Area, Bath and North Wiltshire	112
5. Gloucestershire, Hereford and Worcester	158
6. West Midlands and South Staffordshire	188
7. Shropshire	226
8. Cheshire and North Staffordshire	256
Additional sites	289
General bibliography	292
Name index	295
Subject index	299

Front Cover: The Iron Bridge, Coalbrookdale, Shropshire (Ironbridge Gorge Museum Trust)

Title Page: Menai Suspension Bridge (R. Cragg)

Introduction

This book covers Wales and the western part of central England from Cheshire in the north to just south of Bristol. Areas to the north, east and south are covered in companion volumes in the *Civil Engineering Heritage* series.

Each chapter is related to a specific geographical area and is separately introduced, but in general, the region contains a striking range of landscapes from the wild expanses of the hills of central Wales to the intensively developed industrial areas of the western Midlands, South Wales and the Bristol area. The area is geographically dominated by its major rivers—the Avon, Dee, Severn, Trent, Weaver and Wye—which have formed both a means of communication and a barrier to it. Many of the works described in this book have arisen from these two aspects.

In historical terms the region encompasses the whole range of the industrial development of Britain, starting in the Severn valley around Coalbrookdale where Abraham Darby I, once a Bristol brass founder, made his significant contribution to the rapid advance of the Industrial Revolution from the mid-eighteenth century onwards. From the banks of the Severn industrialization spread rapidly, with the most intensive development taking place in what became known, quite appropriately, as the Black Country. Many 'firsts' will be found, from iron bridges and aqueducts to iron-framed mill buildings and, most recently, Britain's first immersed tube road tunnel at Conwy. Factory-based industrialization requires good transport facilities, both for the movement of raw materials and finished goods. The provision of efficient, low cost transport systems, which began with the unrivalled network of canals in the West Midlands and elsewhere across the region, followed by the intense development of railways, and, in more modern times, the building of the motorway system, has employed the ingenuity of many of Britain's finest engineers.

The names of many of the engineers featured in the text will be familiar to most readers—Thomas Telford, James Brindley, George and Robert Stephenson, Isambard Kingdom Brunel and many others. However, the reader will also find the names of many lesser known engineers, all of whom made their contribution to the development of the profession of civil engineering, and also the names of many of the contractors who undertook the building of the works described. In the past their involvement has often received less credit than is their due.

All the material in this volume has been abstracted from the detailed reports which have been prepared by members of the Institution of Civil Engineers' Panel for Historical Engineering Works and their helpers. These reports are held in the Institution's library and each work is given a Historic Engineering Work (HEW) number. This number, together with the grid reference, is given for each item. Where appropriate, references to sources used are given at the end of each item. A comprehensive subject index is given together with a separate index of engineers, architects and contractors. For those wishing to explore further, a list of additional sites, registered as Historical Engineering Works but not described in detail, is given for each chapter.

In presenting this volume it is hoped that all readers, both those with an engineering background and the more general reader, will obtain a better understanding of the contribution made by civil engineers to the economic and social development of the region and the country as a whole.

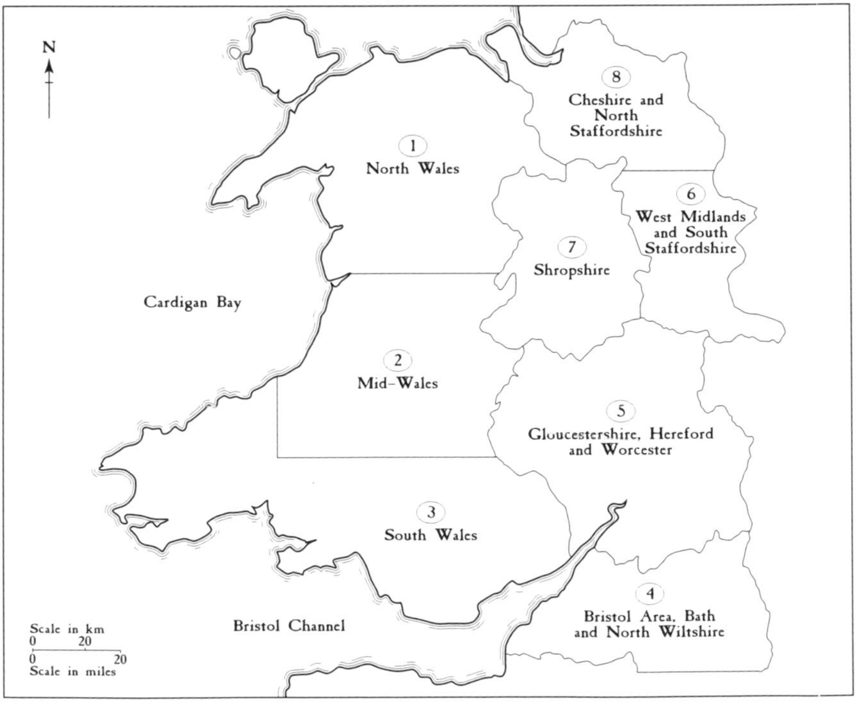

Metric equivalents

Imperial measurements have generally been adopted to give the dimensions of the works described, as this system was used in the design of the great majority of them. Where modern structures have been designed to the metric system, then these units have been used in the text.

The following are the metric equivalents of the Imperial units used:

Length	1 inch = 25.4 millimetres 1 foot = 0.3048 metre 1 yard = 0.9144 metre 1 mile = 1.609 kilometres
Area	1 square inch = 645.2 square millimetres 1 square foot = 0.0929 square metre 1 acre = 0.4047 hectare 1 square mile = 259 hectares
Volume	1 gallon = 4.546 litres 1 million gallons = 4546 cubic metres 1 cubic yard = 0.7646 cubic metre
Mass	1 pound = 0.4536 kilogram 1 Imperial ton = 1.016 tonnes
Power	1 horse power (h.p.) = 0.7457 kilowatt
Pressure	1 pound force per square inch = 0.06895 bar

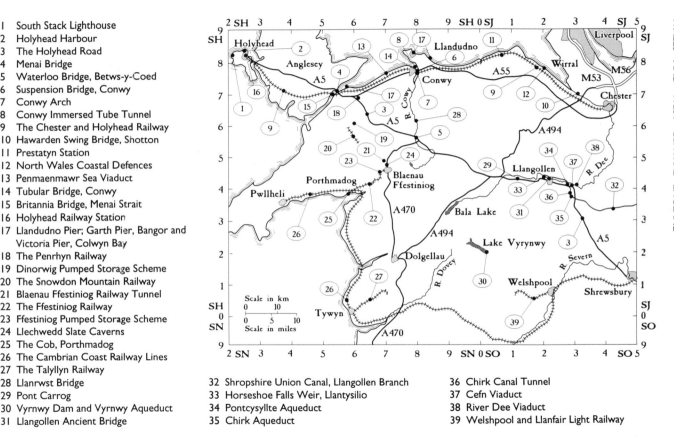

1 South Stack Lighthouse
2 Holyhead Harbour
3 The Holyhead Road
4 Menai Bridge
5 Waterloo Bridge, Betws-y-Coed
6 Suspension Bridge, Conwy
7 Conwy Arch
8 Conwy Immersed Tube Tunnel
9 The Chester and Holyhead Railway
10 Hawarden Swing Bridge, Shotton
11 Prestatyn Station
12 North Wales Coastal Defences
13 Penmaenmawr Sea Viaduct
14 Tubular Bridge, Conwy
15 Britannia Bridge, Menai Strait
16 Holyhead Railway Station
17 Llandudno Pier; Garth Pier, Bangor and Victoria Pier, Colwyn Bay
18 The Penrhyn Railway
19 Dinorwig Pumped Storage Scheme
20 The Snowdon Mountain Railway
21 Blaenau Ffestiniog Railway Tunnel
22 The Ffestiniog Railway
23 Ffestiniog Pumped Storage Scheme
24 Llechwedd Slate Caverns
25 The Cob, Porthmadog
26 The Cambrian Coast Railway Lines
27 The Talyllyn Railway
28 Llanrwst Bridge
29 Pont Carrog
30 Vyrnwy Dam and Vyrnwy Aqueduct
31 Llangollen Ancient Bridge

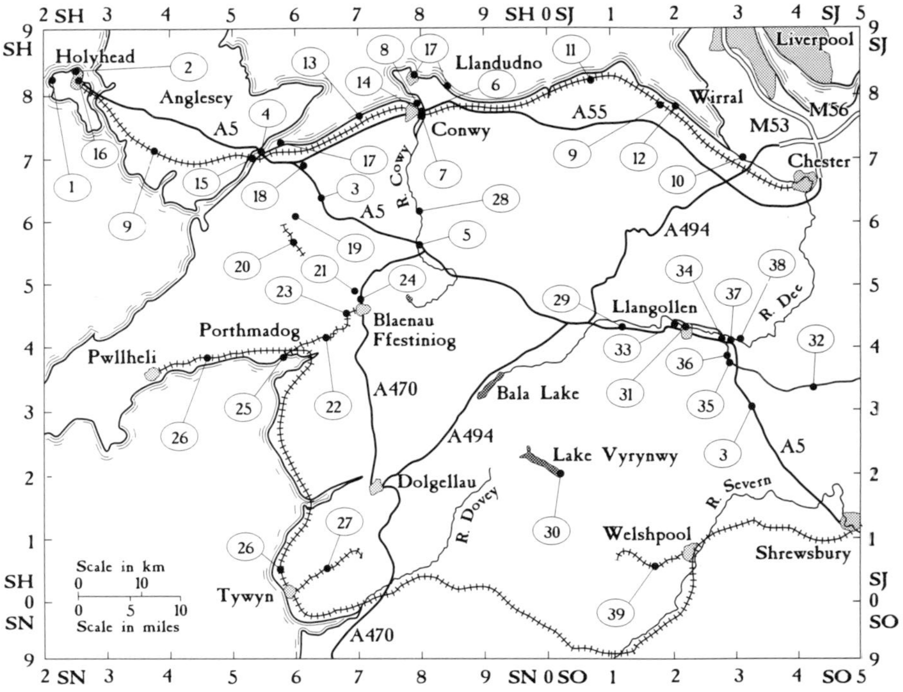

32 Shropshire Union Canal, Llangollen Branch
33 Horseshoe Falls Weir, Llantysilio
34 Pontcysyllte Aqueduct
35 Chirk Aqueduct
36 Chirk Canal Tunnel
37 Cefn Viaduct
38 River Dee Viaduct
39 Welshpool and Llanfair Light Railway

1. North Wales

The boundary between Wales and England commences in the estuary of the River Dee (Afon Dyfrdwy) and skirts Chester less than 2 miles from the city centre. Between Chester and the Vale of Llangollen the political boundary generally follows the Dee valley, but the geographical boundary is the Clwydian Range, which forms its western flank. Beyond the Clwydian Range lie the Vales of Clwyd and Conwy and the natural grandeur of the Cambrian Mountains, which culminate in Snowdon, the highest mountain in England and Wales. Separated by the Menai Strait, the island of Anglesey is still a stronghold of the Welsh language and culture.

The mountainous landscape of North Wales presents serious obstacles to communication and is complemented by many excellent examples of the science and craft of civil engineering, in particular the works of two of the most eminent engineers, Thomas Telford and Robert Stephenson. The route to Ireland via the harbour at Holyhead on Anglesey was of great political importance during the eighteenth and nineteenth centuries following the completion of Telford's Holyhead road. In constructing his highway through the mountains Telford built some of his finest works. Stephenson took an apparently easier coastal route for his railway but this was not without its difficulties and his ingenuity was pushed almost to its limits.

Slate quarrying was an important industry in North Wales until World War II and the steepness of the terrain dictated that narrow gauge rather than standard gauge railways were built to bring the slate down from the mountain quarries to the harbours, whence Welsh slate was exported all over the world. Nowadays the 'great little trains' no longer haul slates, but are a considerable tourist attraction instead.

North Wales has provided industrial England with drinking water and hydroelectricity for many years. The mountains just north of Snowdon form the backdrop for one of the most impressive schemes carried out in this part of Britain during recent years—the hydroelectricity generating station at Dinorwig.

1. South Stack Lighthouse

HEW 756
SH 202 823

The north shore of Anglesey is roughly on the same latitude as Liverpool, rather above the general line of the North Wales coast. The top left-hand corner, as it were, is marked for the main shipping lanes by the Skerries Lighthouse.

Holyhead harbour is on Holy Island, seven miles to the south and is itself marked on the west side by the South Stack lighthouse. In addition it has two local lights, the one on Admiralty Pier being 48 ft high and dating from 1821, the other 63 ft high on the outer breakwater, with both light and breakwater dating from 1873.

South Stack is a tiny rock island of high cliffs cut off by one hundred feet of 'turbulent sea'. Access to it is by 380 steps down to a slender aluminium truss bridge of 110 ft span, which in turn replaced a suspension footbridge of 1828. Before that there was only a breeches buoy arrangement.

South Stack Lighthouse was designed by Daniel Alexander and first lit on 9 February 1809. Built from stone quarried on the site, it consists of a tower of traditional form, tapered and painted white, with gallery and lantern

South Stack Lighthouse

R. CRAGG

about 90 ft above the rock. The lighthouse is flanked by a low long building with a two-span pitched roof and three smaller buildings. These were the engine room and dwellings. The operation is now automatic.

Illumination was originally by oil, then by paraffin vapour and finally, from 1939, by electricity. The light is nearly 200 ft above mean high water and can be seen for over 20 miles.

2. Holyhead Harbour

HEW 1095
SH 250 840

The 63 mile route from Holyhead to Dun Laoghaire is the most direct sea crossing from Great Britain to Dublin.

The history of the port of Holyhead is complicated by the changing names of its elements.

In the seventeenth and eighteenth centuries there was just a creek, which dried out at low water, later known as the Inner Harbour. This was given protection by the Admiralty Pier,[1] designed by John Rennie and completed in 1821, which ran eastward from Salt Island and was linked to the town by a swing bridge over the sound. The Custom House and Harbour Office on Salt Island were part of the scheme. A triumphal arch commemorates the visit of George IV in 1821.

The next stage comprised Telford's Dry Dock[2] of 1825, opposite the tip of Admiralty Pier, and his South Pier, 1831. The Admiralty Pier was then regarded as the North Pier.

The Dry Dock was designed to drain at low water spring tides but by 1829 steam pumping was installed to empty it over the neaps. The Science Museum, London, has a photograph of the Boulton and Watt engine which drove the two bucket pumps. Original drawings suggest that the dock was built with hinged gates, but later a caisson was substituted. The dock is now derelict, although in 1939 it had a new caisson fitted for wartime purposes.

The New Harbour was authorized in 1847, the area within the North and South Piers becoming the Old Harbour.

A breakwater[3] 7860 ft long, built with stone brought down by a railway from quarries on Holyhead Mountain, was under continuous construction from 1845 until 1873. The objective was to enclose a deep water sheltered roadstead of 400 acres, in addition to the 276 acres of harbour. The Engineer was J. M. Rendel, and on his death in 1856 the work was taken over by John Hawkshaw (later Sir John), assisted by Harrison Hayter, who had already been engaged on the project for some years.[4,5]

The Chester and Holyhead Railway came in 1848, first with a terminus at the head of the creek, then in 1856 to the Admiralty Pier which, widened and extended in timber, became known as the Mail Pier, and was used by the Dublin mail-boats from 1849 until 1925. The timber widening and extension were demolished between 1935 and 1942.

Holyhead Harbour

AEROFILMS OF BOREHAMWOOD

Between 1875 and 1880, the London and North Western Railway Company developed the Inner Harbour with two quay walls forming a V which accommodated the new passenger station. A new graving dock was built on the east side of the harbour. In 1922 the channel to the station berths was deepened by a rockbreaker dredger. Previously Telford had had to use a diving bell to trim off submerged rock.

In the past three decades there have been a number of developments in the port, including a deep water wharf for importing aluminium ore, roll-on roll-off berths on the Admiralty Pier and in the Inner Harbour, and a container berth in the latter. The latest development at the port[6,7] is the creation of a terminal facility for a new high-speed catamaran service which can take 1500 people, cars and lorries from Holyhead to Dun Laoghaire in half the time of the previous ferries.

1. Admiralty Pier (HEW 1097) SH 253 829.

2. Old Graving Dock (HEW 1096) SH 254 826.

3. Holyhead Breakwater (HEW 1099) SH 237 837 to SH 258 847.

4. HAWKSHAW SIR J. *Holyhead New Harbour*. Final Report to the Board of Trade, 1873.

5. HAYTER H. Holyhead New Harbour. *Min. Proc. Instn Civ. Engrs*, 1875–76, **44**, 95–130.

6. *Construction News*, 1995, 3 Aug., 13.

7. RUSSELL H. Full steam ahead. *New Civil Engineer*, 1996, 27 Apr., 30–31.

3. Holyhead Road

HEW 1212
SJ 400 179 to
SH 250 832

In the last years of the eighteenth and the first years of the nineteenth centuries the road from London to Holyhead, for the sea crossing to Ireland, was second in importance only to that from London to Dover. But over much of its length it was in a dreadful state of repair, nowhere more so than across North Wales. On Anglesey, reached by ferry, much of the route was only a grass track, and through the Welsh mountains the road ran along the edge of unprotected precipices with gradients as steep as 1 in 6.5. Consequently travel was slow and dangerous: the Irish mail coach took 41 hours to travel from London to Holyhead at an average speed of 6¾ miles per hour.

In 1810 Thomas Telford was commissioned to report

on the state of the road and to suggest improvement, and in 1815 Parliament voted the necessary funds for what was one of the finest achievements of one of the greatest of British civil engineers.[1]

Parliament, through the Holyhead Road Commission, required that the road, which was then under the control of 23 separate turnpike trusts, should be improved throughout its 267 mile length. Telford was appointed to superintend the improvement works. Between London and Shrewsbury the works included the easing of gradients, widening and diversion onto short lengths of new alignment in places. West of Shrewsbury more drastic improvements were necessary and for this section Parliament authorized a Parliamentary Turnpike Commission, which took over the improvement and maintenance of the road from the original turnpike trusts.

By 1819 most of the 85 miles between Shrewsbury and Bangor had been made safe for traffic. Where the road passed through Glyn Diffws, west of Corwen, the hillsides were blasted to enable it to be levelled and widened. From Rhydlanfair on the River Conwy, a 3 mile stretch was built along Dinas Hill with a maximum gradient of 1 in 22. After crossing the Conwy on the Waterloo Bridge into Betws-y-coed, the road followed the south bank of the River Llugwy to a point just upstream of the Swallow Falls, where it crossed the river and then took up a new alignment for 1 mile before rejoining the old line, and so on to Capel Curig. A new alignment through the Nant Ffrancon Pass[2] led through Bethesda to Bangor. Between 1820 and 1828 Telford built 20 miles of new road across Anglesey from Menai Bridge to Holyhead, including the Stanley Embankment[3] between Anglesey and Holy Island.

This section of Telford's route is now the modern A5 trunk road. Although the turnpike system lapsed in the 1880s there remain here and there, as in other parts of the country, relics in the shape of toll-houses and road furniture. Among the former are those at Caergiliog and Llanfairpwllgwyngyll. Among the latter are the distinctive 'rising sun' pattern wrought iron toll-gate and the milestone[4] at the western end of the Stanley Embankment. Examples of a Telford toll-house, toll-gate and milestone can also be seen at the Blists Hill Museum, Ironbridge (SJ 695 032).

Other sections of the Holyhead Road are described in Chapter 6, which also includes a description of Telford's road construction methods, and Chapter 7.

1. TELFORD T. Reports to the Commissioners for the road from London to Holyhead, 1824–34. (Copy in Institution of Civil Engineers library.)

2. Nant Ffrancon Pass (HEW 460) SH 722 582 to SH 626 659.

3. Stanley Embankment (HEW 1246) SH 285 798 to SH 276 802.

4. Telford Milestone (HEW 459) SH 276 803.

4. Menai Bridge

HEW 109
SH 556 715

In 1817 the Holyhead Road Commissioners instructed Telford to prepare plans for a bridge to replace the ferry across the Menai Strait. The plans, for what is generally regarded as Telford's finest work,[1] were ready by February of the following year.

The construction took John Wilson, the contractor, a period of seven years from 1819. The bridge has an overall length of 1000 ft, seven stone approach spans of 52 ft and a main central suspension span of 579 ft, tower to tower, carrying the road 100 ft above sea level.

The piers are faced with Anglesey marble. The deck was suspended from four sets of wrought iron chains, the links for which were made by William Hazledine at Upton Forge near Shrewsbury and tested in his Coleham workshops.

Modifications to the bridge were made following damage by a storm in January 1839.[2,3] In 1893 Sir Benjamin Baker replaced the timber deck by steel troughing on flat-bottomed rails. In 1940 the chains were replaced by two sets of steel chains, the deck was rebuilt in steel to take heavier road traffic and a cantilevered footway was added on each side, so giving the bridge the appearance that it has today.[4] Nevertheless, the modern alterations do not detract from the gracefulness of Telford's original structure, set as it is in magnificent scenery.[5]

A very full account of the design and construction is given by W. A. Provis, the Resident Engineer, in a large folio volume published in 1828.[6]

1. PAXTON R. A. Menai Bridge (1818–1826) and its influence on suspension bridge development. *Trans. Newcomen Soc.*, 1977–78, **49**, 87–110.

2. MAUDE T. J. Account of the alterations made on the structure of the Menai Bridge, during the repairs in consequence of the damage it received from the gale of January 7, 1839. *Trans. Instn Civ. Engrs*, 1842, **3**, 371–375.

3. PROVIS W. A. Observations on the effects of wind on the suspension bridge over the Menai Strait, more especially as relates to the injuries sustained by the roadways during the storm of January, 1839. *Trans. Instn Civ. Engrs*, 1842, **3**, 357–370.

4. MAUNSELL G. A. Menai Bridge reconstruction. *J. Instn Civ. Engrs*, 1946, **25**, No. 3, Jan., 165–206.

5. HUSBAND J. The aesthetic treatment of bridge structures. *Min. Proc. Instn Civ. Engrs*, 1901, **145**, 166.

6. PROVIS W. A. *An historical and descriptive account of the suspension bridge constructed over the Menai Strait in North Wales.* Ibbotson and Palmer, London, 1828.

5. Waterloo Bridge, Betws-y-coed

HEW 106
SH 798 557

This well-known bridge was built by Telford in 1815 to carry the London to Holyhead road over the Afon Conwy. Its single span of 105 ft consists of five cast-iron arched girders at 5 ft centres supporting cast-iron deck plates. The outer ribs carry the legend 'This arch was constructed in the same year the battle of Waterloo was fought', and the spandrels above are beautifully decorated with rose, thistle, shamrock and leek, modelled in relief by William Hazledine, whose foreman, William Stuttle, was responsible for the erection of the bridge.

In 1923 the bridge was strengthened by concreting the inner three ribs and adding a 7 in. reinforced concrete deck. This was cantilevered to provide new footways and allow the roadway to be widened.

In 1978 a new 10 in. reinforced concrete deck was added and the original cast-iron parapet fence, visible from outside, was protected by an additional fence in-

Waterloo Bridge

INSTITUTION OF CIVIL ENGINEERS

side. The masonry abutments were strengthened on both occasions.

6. Suspension Bridge, Conwy

HEW 107
SH 785 776

Opened in 1826, this was the most successful of Thomas Telford's six 'Gothic' bridges. Two pairs of solid ashlar limestone towers 12 ft 4 in. in diameter and 40 ft high support the chains, which span 327 ft between their supports on the towers. They are linked by castellated walls containing the 10 ft wide carriageway arches. The wrought iron chains are arranged in two tiers of five 9 ft long links joined by deeper plates, the joints of which are alternated. Vertical rods at 5 ft intervals carry the deck suspended from the junction plates. The present deck is a replacement dating from 1896. The original deck was probably made up of a light iron framework braced by bars on its underside, upon which were laid two longitudinal layers of fir planks. Plans to demolish the bridge following the construction of a new road bridge alongside led to a world outcry in 1958, since when the bridge has been in the care of the National Trust and closed to vehicular traffic.

Conwy Bridges: tubular bridge, suspension bridge and steel arch bridge

AEROFILMS OF BOREHAMWOOD

7. Conwy Arch

HEW 692
SH 785 775

The 310 ft span steel arch at Conwy, built in 1958 to relieve Telford's suspension bridge of the weight of road traffic, is a very fair solution to the problem of blending a modern structure into a pleasing environment, which in this case is dominated by Conwy Castle and by the Telford and Stephenson bridges immediately upstream.

It was intended to build a bridge twice as wide and the present bridge is only the first half. The two sides differ. The upstream side shows a simple spandrel-braced arch of N-type and a short approach viaduct of short spans on columns.

The castle and bridges are mostly viewed from downstream, from which direction the viaduct is masked by what appears to be a long wing wall with a pilaster resembling an old type cutwater and proportioned like the turret of a castle. From this springs a flat and graceful arch which, combined with the vertical flat curve of the roadway, gives the general impression of an ancient road bridge. The castellated towers of the other two bridges can just be seen in the background and the simplicity of the newer bridge makes a satisfactory contrast enhancing both, as can be seen from the coloured tourist postcards.

8. Conwy Immersed Tube Tunnel

HEW 2069
SH 776 786 to
SH 787 782

The latest crossing of the Conwy, by the new A55 road, uses a very different technique.[1,2,3] After much discussion, it was decided to build Britain's first immersed tube tunnel, sited downstream of the earlier crossings. The complete tunnel includes an eastern 'cut and cover' tunnel 260 m long, a centre section of immersed tube under the river 710 m long and a final western 'cut and cover' section of 120 m, giving a total length of 1090 m.

The immersed tube section is formed of six large prefabricated reinforced concrete units, each 118 m long, 24.1 m wide and 10.5 m high, weighing 30 000 tonnes. Each unit was cast in a coffer-dam area on the west side of the estuary and floated out into position in the river before being sunk into its final position on the river bed.

Conwy Immersed Tube Tunnel: west portal

All the six units were then joined together and the roadway completed.

The £102 million contract was let to Costain Tarmac Joint Venture in 1986 and work started in November of that year. The Engineers for the scheme were Travers Morgan and Partners in association with Christiani and Nielsen A/S. The Conwy submerged tube crossing was opened by Her Majesty the Queen on 25 October 1991.[4]

1. DAVIES G. W. *et al.* The Conwy tunnel—scheme development and advanced works. *Immersed tunnel techniques. Proc. Conf. Instn Civ. Engrs, Manchester, April 1989.* Thomas Telford, London, 1990, 125–144.

2. STONE P. A. *et al.* The Conwy tunnel—detailed design. Op. cit., 277–300.

3. MCFADZEAN J. F. *et al.* The Conwy estuary tunnel—construction. Op. cit., 357–372.

4. RUSSELL L. (ed.) A55 project special. *New Civil Engineer,* 1994, Aug.

9. Chester and Holyhead Railway

HEW 1094
SJ 413 670 to SH 248 822

The success of the Liverpool and Manchester Railway[1] showed that rail travel offered speeds three times those possible on the roads and effectively made Telford's

Robert Stephenson, Engineer of the Chester and Holyhead Railway

excellent improvements to the Holyhead Road out of date almost as soon as they were completed. Communications between England and Ireland remained important. Of the three railway schemes considered for this area, the one which was adopted was that of the Chester and Holyhead Railway, incorporated in 1844 with Robert Stephenson as Engineer-in-Chief and completed in 1850 over a route surveyed by his father in 1838. Alexander Mackenzie Ross was Stephenson's assistant and among the main contractors were Edward Betts, William Mackenzie, Thomas Brassey, John Stephenson and Thomas Jackson. Later, in 1847, with the opening of the line in prospect, Hedworth Lee, who had been Ross's assistant, was appointed Resident Engineer for the railway.[2,3]

The line skirts the North Wales coast and for some 43 miles the special feature of the route was the need for heavy engineering works to protect the railway and coast from the sea. The new A55 road along the coast faced similar problems, which modern engineers have dealt with by adopting solutions similar to those of Stephenson, although some 22 000 precast concrete 'dolosse' units were used to protect the embankment at Llandulas.

Important viaducts on the railway are the 22-arch Ogwen Viaduct[4] at Talybont, just north of which is Penrhyn Castle, and that at Malltreath[5] on Anglesey, but of prime importance are the river crossing at Conwy and the Britannia Bridge over the Menai Strait.

Near its western terminus, the railway crosses the strait between Anglesey and Holy Island alongside Telford's Stanley Embankment.[6]

Francis Thompson designed the original stations as two-storey rectangular structures in the Georgian style. Of these, one of the smallest, which is the first reached after crossing the Menai Strait, is the best known to tourists, solely because of the length of its name—Llanfairpwllgwyngyllgogerychwyrndrobwll-llantysiliogogogoch—popularly abbreviated to Llanfair P.G.

The Chester and Holyhead Railway was built for speed, a feature emphasized in 1857 when John Ramsbottom installed the first ever water troughs near Mochdre, later re-sited near Aber (SH 652 724). The troughs were several hundred yards long and located between the rails. By lowering a scoop into the trough the locomotives picked up water and this made non-stop long-distance rail travel a reality by avoiding the necessity for trains to stop for rewatering.

1. Rennison R. W. *Civil engineering heritage: Northern England*. Thomas Telford, London, 1996, 249–255 (HEW 223).

2. Parry E. *Railway companion from Chester to Holyhead*. Catherall, Chester, 1849, 2nd edn; facsimile by E. and W. Books, London, 1970.

3. Baughan P. E. *The Chester and Holyhead Railway*. David and Charles, Newton Abbot, 1972, 56–58, 77–78.

4. Ogwen Viaduct, Talybont (HEW 1280) SH 602 707.

5. Malltreath Viaduct, Anglesey (HEW 1288) SH 414 691.

6. Stanley Embankment (HEW 1246) SH 276 802 to SH 285 798.

10. Hawarden Swing Bridge, Shotton

HEW 775
SJ 312 694

Three spans of steel hogback N trusses carry two railway tracks over the River Dee. The two fixed spans are each 125 ft and the swing section is of a record length of 287 ft, which gave a clear opening of 140 ft. Movement was originally by hydraulic rams and chains to a 32 ft diameter circular girder. The 90 ft landward end of the bridge ran on a quadrant rail. The bridge was fixed and the machinery removed in 1971. The girders have a maximum depth of 32 ft and are at 27 ft 6 in. centres. There are cross girders at 17 ft centres with two pairs of rail girder bearers.

The bridge was built in 1887–89 by John Cochrane and Sons for the Manchester, Sheffield and Lincolnshire Railway (later part of the Great Central Railway), which, in spite of its name, had a line as far into Wales as Wrexham.

Francis Fox[1] was the Engineer and the bridge was opened on 3 August 1889 by the wife of W. E. Gladstone, after whose residence it was named.[2]

1. FOX F. The Hawarden Bridge. *Min. Proc. Instn Civ. Engrs*, 1892, **108**, 304–317.

2. Hawarden Dee railway bridge. *Ill. London News*, 1889, 10 Aug., 184.

Hawarden Swing Bridge

R. CRAGG

R. CRAGG

Prestatyn station

11. Prestatyn Station

HEW 1208
SJ 064 831

Prestatyn station, on the line of the Chester and Holyhead Railway, is one of the few remaining examples of the prefabricated single-storey station buildings manufactured at the Crewe works of the London and North Western Railway in the closing years of the nineteenth century.

The buildings were constructed in 6 ft 8 in. modules and the timber framework was faced externally with rusticated boarding. Brick footings, fireplaces and chimneys were used where required. Canopies with valances were cantilevered over the platforms, their timber beams being supported on iron brackets. Prestatyn station was rebuilt in the original style in 1979.

12. North Wales Coastal Defences

HEW 1228
SJ 20 78 to SH 68 76

About one-half of the entire length of the Chester and Holyhead Railway is beset with unending maintenance problems arising from river and tidal flooding, coastal erosion and storm damage. As a result, numerous important engineering works have had to be built.

At Holywell 2½ miles of embankment have constantly to be replenished by tipping slag some 100 yd from the railway. There is another wall 3900 yd long between Mostyn and Point of Ayr.

The Abergele railway embankment, known as The Cob, consists of 1960 yd of shingle beach and groynes flanked by the 1470 yd Rhuddlan Marsh wall, first embanked in 1800 by trustees and raised and improved when the London and North Western Railway took it over in 1880.

At Llysfaen, east of Penmaenrhos Tunnel, landslips have frequently occurred, only partly compensated for until now by the fact that they have produced material to replenish the beaches to the east. Currently the problem here has in effect been transferred from rail to road. The A55 dual carriageway[1,2] has been built to seaward of the railway and is protected by some 22 000 'dolosse' units: these are specially shaped interlocking concrete objects, weighing 5 tons each and placed on a 1 in 2 slope as facing to secondary armour of one ton rockfill, with a smaller rock core and a filter membrane.

At Old Colwyn there is an extensive area of unstable ground below the railway. For 680 yd there are groynes, toe walls, larger walls, pitched slopes and wave breakers.

From Colwyn Bay to Conwy, protection is given by the promontory that terminates in the Great Ormes Head.

Further west, as far as Llanfairfechan, the problem is different. High cliffs forced the coach roads of Telford's day and earlier, and later Stephenson's railway, to make use of shelves, either on sea walls at the base or higher up on the cliffs themselves, and occasionally to burrow through headlands in tunnels.

East of Penmaenmawr half a mile of railway is protected by a 20 ft wall leading to Penmaenbach Tunnel. Part of this wall was destroyed in 1945 and was replaced to a new design in reinforced concrete.

The 253 yd Penmaenmawr Tunnel is extended by avalanche shelters to guard against rockfalls from the cliffs towering above it.

The Penmaenmawr Sea Viaduct is part of a 1500 yd length of solid masonry sea wall up to 40 ft high, located some 20 ft above high water level and surmounting a 1 in 30 pitched slope down to the beach.

Coastal defence is an important branch of civil engineering, often dramatic and always expensive. However, failure of coastal defences can be even more expensive, as was demonstrated during February 1990 when a section of the defences failed at Towyn, flooding the town to a considerable depth.

Although not a work of coastal defence, it is worth recording here that along this section of the Welsh coast, at Rhyl, took place what must be one of Britain's earliest river navigation works. In about 1277, as part of the work being carried out to refurbish Rhuddlan Castle for King Henry III, the section of the River Clwyd between Rhuddlan and the sea was rendered navigable for seagoing ships.[3] This was achieved by straightening the meandering course of the river over a length of some 2¾ miles. To carry out this work, diggers were recruited from Lincolnshire, some 300 being brought to Rhuddlan under mounted guard. The work was carried out between 1277 and 1280, the wages of the diggers employed on the river works being recorded in the castle accounts.

1. PARKINSON J. A55 coast road raft floats on sea of piles. *New Civil Engineer*, 1983, 27 Jan., 20.

2. GREEMAN A. Coast road treads softly through Colwyn Bay. *New Civil Engineer*, 1984, 9 Feb., 30–32 and cover.

3. BROWN R. A. *et al. The history of the King's Works, Volume 1: the Middle Ages*. HMSO, London, 1963, 318–321.

13. Penmaenmawr Sea Viaduct

HEW 829
SH 696 762

On 22 October 1846, during the construction of the sea wall at Penmaenmawr to protect the Chester and Holyhead Railway, a severe storm with 40 ft waves destroyed the most exposed section of the wall in the presence of the Engineer, Robert Stephenson.

He decided to replace it by an open viaduct, some 182 yd long, of 13 equal spans through which the waves could dissipate their energy up the sloping beach beneath.

The masonry piers, which were built in coffer-dams, are 32 ft wide and 6 ft 4 in. thick and stand to a height of 41 ft, about 15 ft above high water. They were protected from scour by piles retaining large boulders set in concrete to form a pavement extending some 18 ft seaward.

The original deck consisted of four longitudinal cast-iron girders under each track, of inverted T-section, 24 in. deep, 14 in. wide at the bottom and 4½ in. wide at the top, carrying timber way beams. The decking was of timber laid crosswise.

Work began in March 1847 and the viaduct opened to traffic on 1 May 1848. The contractor was Wharton and Warden.

In 1908 the deck system was replaced by six-ring brick arches of 13 ft rise.

14. Tubular Bridge, Conwy

HEW 108
SH 785 774

This bridge was built in 1848 by Robert Stephenson to carry the Chester and Holyhead Railway across the River Conwy. It is very similar in design to his other great bridge on the same line, the Britannia Bridge and, like it, marked a significant advance in the art of bridge building.[1] It consists of a single span 400 ft long, formed by two parallel rectangular wrought iron tubes, each weighing 1300 tons. These were built ashore and then floated out on pontoons to be raised in position onto stone abutments on either side of the river. Masonry towers were built on the abutments and topped with battlements and turrets to harmonize with the adjacent Conwy Castle.

Before the bridge was commissioned, loading tests were carried out on both tubes using up to 300 tons of iron kentledge, which produced a central deflection of 3 in. By way of comparison, the passage of an ordinary train is said to produce a deflection of only ⅛ in.

Two piers added in 1899 reduced the span by 90 ft.

The railway approaches the bridge beside the embankment built by Telford which leads to his elegant suspension bridge.

1. CLARK E. *Britannia and Conway tubular bridges*. Day, London, 1850.

15. Britannia Bridge, Menai Strait

HEW 110
SH 542 710

Opened in 1850 to carry the Chester and Holyhead Railway across the Menai Strait, the Britannia tubular bridge,

INSTITUTION OF CIVIL ENGINEERS

Britannia Bridge, Menai Strait: Robert Stephenson is seated in the centre and Joseph Locke and I. K. Brunel are seated on the right

together with that at Conwy, may be said to be the forerunner of the box girders of today.

Robert Stephenson contemplated using suspension spans with deep rectangular trough-shaped stiffening girders, but was struck with the idea that if the top of the trough was closed in, the girder might be self-supporting. Studies by Professor Eaton Hodgkinson into the strength of materials and careful testing of large models by the structural engineer and shipbuilder William Fairbairn confirmed the feasibility of this approach.[1]

As at Conwy, the tracks were carried within twin rectangular riveted tubes built up from wrought iron plates, as developed for shipbuilding. The two main spans were each of 460 ft, flanked by side spans of 230 ft. With spans greatly in excess of any previously constructed, the successful building of the Britannia and Conwy bridges was an outstanding advance in the use of wrought iron in structures.

The four tubes for the two mainstream spans, each weighing 1800 tons, were built on the Caernarfon shore, then floated out and jacked up 100 ft onto the towers. The same procedure had been used at Conwy and was subsequently adopted by I. K. Brunel for his Chepstow and Saltash bridges.[2] In June 1849, Brunel stood beside his great friend and rival Stephenson to watch the launching of the first Menai tube.[3]

All four spans for each track were connected end to end through the towers to form a 1511 ft long continuous girder and so take advantage of the economy of material stemming from continuity. With its adherence to straight lines, its massive masonry abutments and its three masonry towers rising over 100 ft above the bridge, the structure presented an 'outstanding example of symmetry, harmony and proportion'.[4]

In 1970 a fire destroyed the protective timber roof above the tubes and the heat caused the tubes to tear apart into four separate spans, the effect of which was greatly to reduce the strength of the bridge and to render it unusable. It was replaced by a clever, practical and modern structure of quite different design,[5] which still retains Stephenson's original stone towers complete with the monumental lions at the ends of the bridge. The new main spans, designed by Husband and Company of Sheffield, are steel arches with eight panels of N truss spandrel bracing in each half arch—an arrangement very similar to that of the longer Victoria Bridge built some 70 years earlier over the Zambesi River between Zimbabwe and Zambia.[6] Each side span was divided into three spans built in reinforced concrete. Both bridges were built by the Cleveland Bridge and Engineering Company of Darlington. The new Britannia Bridge carries above the rail tracks a road onto which the A5 trunk road was diverted, so that Telford's Menai Bridge was relieved of heavy traffic. A section of one of the original box girders has been preserved as a monument nearby, with an explanatory notice beside it.

1 Fairbairn W. Account of the construction of the Britannia and Conway tubular bridges. *Min. Proc. Instn Civ. Engrs*, 1850, **9**, 233–287.

2. Otter R. A. *Civil engineering heritage: Southern England.* Thomas Telford, London, 1994, 41–42 (HEW 29).

3. Seyrig T. The different modes of erecting iron bridges. *Min. Proc. Instn Civ. Engrs*, 1880–81, **63**, 161–162.

4. FOWLER C. E. *The ideals of engineering architecture*. Spon, London, 1929, 171.

5. HUSBAND H. C. *et al*. Reconstruction of the Britannia Bridge. *Proc. Instn Civ. Engrs*, 1975, **58**, Feb., 25–66.

6. HOBSON G. A. The Victoria Falls Bridge. *Min. Proc. Instn Civ. Engrs*, 1907, **170**, 1–23.

16. Holyhead Railway Station

HEW 1098
SH 248 822

The Chester and Holyhead Railway first opened a station at Holyhead harbour in 1848.[1] This was a temporary station and was replaced by a second station nearer the harbour in 1851. Finally, in conjunction with the further development of the Inner Harbour by the London and North Western Railway (L&NWR), a large new station and hotel were built, opened by HRH the Prince of Wales in June 1880.

The layout was a V shape, with the down side track and buildings along the west arm of the harbour and the up along the east arm, with the hotel at the south end in the V. Of the original overall roof structures, only the up side remains. It is a good example of the type of overall roof used by the L&NWR at that time for several of their principal stations, as at Huddersfield in 1885.[2]

Holyhead Station Roof

R. CRAGG

The site is curved and tapering and the 64 roof trusses vary in span from 62 ft downwards. They are supported on a substantial screen wall 25 ft high on the one side and on cast-iron columns with short wrought iron lattice girders between, on the other.

With a 1 in 2 roof slope and a camber of about a foot in the main tie, the trusses are of distinctive design, with a clerestory carrying louvred ventilators. The main queen posts are braced by four tie rods and a ring.

1. BAUGHAN P. E. *The Chester and Holyhead Railway*. David and Charles, Newton Abbot, 1972, 228–233, 242–243.

2. RENNISON R. W. *Civil engineering heritage: Northern England*. Thomas Telford, London, 1996, 188 (HEW 446).

17. Llandudno Pier; Garth Pier, Bangor; Victoria Pier, Colwyn Bay

HEW 432
SH 784 830

Llandudno Pier[1,2] was designed by James Brunlees and Alexander McKerrow and built by John Dixon in 1876–77. The 2295 ft long structure is in two sections. The main pier is carried on a wrought iron lattice-girder framework supported on cast-iron columns and extends 1234 ft to a T-shaped pierhead 60 ft wide. The deck is lined with four pairs of kiosks, with three larger kiosks at the head. At

Llandudno Pier

R. CRAGG

the shore end, an arm of the platform connected with an 1883–84 pavilion some distance along the shore. The pavilion had a projecting gable centre portion and recessed wings with apsidal ends, and was fronted by a verandah. The pavilion, which housed in its basement what was at one time the largest indoor swimming pool in Britain, was burned down in February 1994.

Garth Pier at Bangor was designed by John James Webster and built in 1896 by Alfred Thorne. This 1500 ft promenading and embarkation pier was founded on cast-iron screw piles, and its deck was supported by steel girders, which had to be replaced during the 1980s.

HEW 427
SH 584 732

Victoria Pier at Colwyn Bay was designed by Maynall and Littlewood and was built by the Widnes Foundry Company in 1899–1900, also using cast-iron screw piles. Another comparatively late pier, this also used steel in its deck construction. The original pavilion was destroyed by fire in 1923, after which the length of the pier was increased to 475 ft and a new pavilion was opened in 1934.

HEW 1285
SH 853 792

1. DUNKERLEY P. Seaside piers. *Research day school: maritime engineering around the Irish Sea*. Merseyside Museum, Liverpool, 19 Feb. 1994.

2. ADAMSON S. H. *Seaside piers*. Batsford, London, 1977, 40–46, 54, 72–87, 93, 102, 104, 113.

18. Penrhyn Railway

HEW 1257
SH 592 730 to
SH 616 660

The Cegin Viaduct on the Chester and Holyhead Railway crosses the line of the narrow gauge Penrhyn Railway, which used to carry slate to Port Penrhyn from the quarries at Cefn-y-Parc near Bethesda some 6½ miles inland. The quarries were in use as early as 1580 and were served by packhorses. In 1782 large scale working began and the roads were improved. In 1801 packhorses were superseded by a horse waggonway, one of the few in Wales to use edge rails and flanged wheels. The first rails were unusual, being of oval section. The line was rebuilt to 1 ft 10¾ in. gauge in 1874 with steam locomotives, as many as 28 being at work at one time, mostly in the quarry itself, which is now served by road. From 1879 until 1958 the line carried workmen as passengers. It finally closed in 1962.[1]

An excellent selection of relics from the railway, of which a few traces remain on the ground, is preserved in the Industrial Railway Museum, opened in 1965 in the stable block of Penrhyn Castle (SH 603 720).[2] Some of the permanent way point and crossing items are particularly interesting. One of the narrow gauge locomotives on display is the 0–4–0T Hunslet *Charles* of 1882, whose sisters *Linda* and *Blanche*, of 1893, are now in use on the Ffestiniog Railway.

Penrhyn Castle itself was designed by Thomas Hopper for Mr G. H. D. Pennant and built between 1827 and 1837. It now belongs to the National Trust.

1. BAUGHAN P. E. *A regional history of the railways of Great Britain, Volume 11: North and Mid Wales.* David and Charles, Newton Abbot, 1980, 110–111.

2. SIMMONS J. *Transport museums in Britain and Western Europe.* Allen and Unwin, London, 1970, 141.

19. Dinorwig Pumped Storage Scheme

HEW 1236
SH 598 607

On the opposite side of Llyn Peris from the lower terminus of the Snowdon Mountain Railway is the site of the largest pumped storage power station in Europe and the third largest in the world.[1]

As the name implies, pumped storage is in effect a means of storing energy. Electricity generated from base-load power stations at periods of low demand is used to pump water from one reservoir to another at a higher level. At periods of peak demand, water is released to the lower reservoir, passing on its way through turbines which drive electric generators feeding to the national grid. This avoids the necessity for providing expensive thermal power stations solely to deal with peak loads, or for operating base-load stations at below their economic capacity in order that output can be increased to meet fluctuations in demand. Feeding the pumping mode of the pumped storage scheme permits more constant and therefore more economic operation of thermal base-load stations. Hydroelectric stations can be brought on power in seconds, which makes this system invaluable for dealing with sudden demands for electricity such as occur at the end of popular television programmes or on the occasion of unexpected breakdowns on the grid.

The upper reservoir at Dinorwig is the existing lake of Marchlyn Mawr,[2] the capacity of which has been raised by a 1970 ft long rockfill dam 118 ft high. This permits the water to rise 108 ft above the original level to accommodate the operation fluctuation of 100 ft. The upstream face of the dam is sealed by a layer of asphaltic concrete, only the second instance of such an application in Britain.

From Marchlyn Mawr, 5575 ft of 34½ ft diameter low pressure tunnel lead on a slight gradient to a 33 ft diame-

Dinorwig Pumped Storage Scheme: power station machine hall under construction

ter vertical shaft 1475 ft deep. From the foot of this shaft, a 31 ft diameter high pressure tunnel leads to the power station about 2200 ft away, dividing into six smaller diameter tunnels before reaching the turbines.

The machinery is housed in nine man-made caverns under the Elidir Mountain, the largest, the main machinery hall, being 590 ft long, 80 ft wide and 197 ft high. The nearby transformer hall is 530 ft long, 80 ft wide and 62 ft high; there are other massive shafts and galleries for hydraulic and electrical control equipment.

The six turbines work in reverse as pumps with their alternators as motors. On average, pumping lasts six hours each night and generation five hours each day, with an average output of 1680 MW at 18 000 V from an installed capacity of 1800 MW. A load of up to 1320 MW can be picked up in 10 seconds. From the turbines, six 12 ft diameter tunnels lead in pairs to three 27 ft diameter tail race tunnels which discharge into Llyn Peris, some 2000 ft away and 1650 ft below Marchlyn Mawr.

The project was designed by the former Central Electricity Generating Board with James Williamson of Glasgow, in association with Binnie and Partners of London, as consultants. When the main underground works were let in November 1975, a record was set for the highest value civil engineering contract ever let in the United Kingdom.

The station was fully commissioned at the end of 1983 and formally opened by HRH the Prince of Wales on 9 May 1984.

Great care was taken to minimize the effect of the works on the environment, in the heart of Snowdonia. Almost all of the construction is underground: even the 400 kV outgoing transmission cables are buried for a distance of 6 miles. The few administrative buildings above ground are of local stone, much of it from old quarry buildings. A good deal of the spoil from the excavations, which include a total of 10 miles of tunnels, together with heaps of slate waste which had disfigured the neighbourhood for years, was tipped into old quarries or the deeper parts of Llyn Peris.

A short distance away at Llanberis is the water-wheel[3] which drove the machinery for the Dinorwig slate works, now the Welsh Slate Museum and part of the National Museum of Wales. With its 50 ft 6 in. diameter, it is one of the largest water-wheels in the United Kingdom.

1. BAINES J. A. *et al.* Dinorwig pumped storage scheme. *Proc. Instn Civ. Engrs*, 1983, **74**, Nov., 635–718.

2. Marchlyn Mawr Reservoir (HEW 1237) SH 617 619.

3. Llanberis Waterwheel (HEW 1284) SH 585 602.

20. Snowdon Mountain Railway

HEW 1222
SH 582 597 to SH 609 543

The Snowdon Mountain Railway has the double distinction of being the only rack railway in Britain and having the highest station, at 3493 ft above sea level. It was designed by Douglas and Francis Fox, using the Swiss system developed by Dr Roman Abt, and was built by A. H. Holme and C. W. King of Liverpool between December 1894 and January 1896. Their men worked a five day week, unusual in those days, because of the arduous conditions on the mountainside. The line rises 3140 ft, in a distance of 4 miles 1100 yd from Llanberis to the summit station on Yr Wyddfa, the highest peak on Snowdon. The average gradient is 1 in 7.8, with the steepest being 1 in 5.5 and the mildest 1 in 50. The line was designed to the European narrow gauge of 800 mm (2 ft 7½ in.) for the

R. CRAGG

Snowdon Mountain Railway

running rails, which are attached to steel sleepers. The two central rack blades are bolted either side of steel chairs, which are in turn bolted centrally to each sleeper. The line is single track with passing loops at the stations.

There are ten bridges, the largest being the Lower Viaduct over Afon Hwch with 13 arches spanning 30 ft and one skew arch spanning 38 ft 4 in. over a road. This structure is 500 ft long and built on a gradient of 1 in 8.5.

21. Blaenau Ffestiniog Railway Tunnel

HEW 755
SH 688 503 to SH 697 469

This 2 mile 340 yd long tunnel under 1712 ft high Moel Drynogydd is the seventh longest railway tunnel in Britain.[1] It was designed initially by Hedworth Lee of the London and North Western Railway (L&NWR) to take a narrow gauge single-track line from Betws-y-coed to Blaenau Ffestiniog to join up with the Ffestiniog Railway.[2] Construction began on 6 December 1873 with the contractor Gethin Jones starting work on the northern of the three shafts used for the excavation of the tunnel. Eight headings were used from the bases of the shafts and the portals, and extremely hard rock was encountered, causing Jones to abandon the work. The tunnel was completed by direct labour under the supervision of William Smith, who succeeded Lee as District Engineer at Bangor.

During construction, the L&NWR decided to alter the line to standard gauge—although this meant dispensing with the connection to the Ffestiniog Railway—and altering the work already done. The tunnel is 18 ft 6 in. high and 16 ft 6 in. wide and is straight except for short curved sections at either end. Most of the tunnel is on an ascending or descending gradient of 1 in 660 except for a short level at the summit near the southern end of the tunnel, where the rail level is 673 ft above that at Betws-y-coed station. The tunnel was opened to traffic on 22 July 1879.

On the same line is the Lledr Viaduct,[3] crossing the A470 Afon Lledr road, an occupation road and woodland. It is popularly known as 'Gethin's Bridge' after Gethin Jones, the contractor who built it.

1. SMITH W. Summit-level tunnel of the Betws and Ffestiniog Railway. *Min. Proc. Instn Civ. Engrs*, 1882–83, **73**, 150–177.

2. DUNN J. M. From Llandudno Junction to Blaenau Ffestiniog. *Rly Magazine*, 1959, **105**, Dec., 821–822.

3. Lledr Viaduct (HEW 1303) SH 780 539.

22. Ffestiniog Railway

HEW 647
SH 571 384 to
SH 701 459

This 1 ft 11½ in. gauge railway was designed by James Spooner and his two sons and was opened in 1836 to take slate from the Blaenau Ffestiniog quarries to the harbour at Porthmadog.[1,2] It descended 700 ft in 13 miles, the maximum gradient being 1 in 80. Upgoing trains were drawn by horses for most of its length, except on inclines which were redesigned by Robert Stephenson with waterwheels to provide the traction. Downhill the horses rode in a special car with the loaded wagons going down under gravity.

The line crosses the bay at Porthmadog on the Cob. The carefully graded route necessitated embankments up to 60 ft high, built of slate with almost vertical sides, and two tunnels at Garnedd and Moelwyn. Just west of Tan-y-Bwlch station the line crosses the B4410 on a neat little cast-iron arch bridge[3] of 18 ft skew and 12 ft square span, built in 1854.

Steam traction was introduced in 1863 by James Spooner's son Charles and passenger traffic began in 1864. The first bogie rolling stock in Britain was introduced on this line. The line was closed in 1946 but restoration work was begun by the Ffestiniog Railway Society, many of whose members rebuilt the railway in their own time to make it a highly successful tourist attraction.[4] The construction of the Ffestiniog power station meant diverting part of the line. The diversion begins at SH 679 422 just beyond Dduallt station, where it turns off to the east and makes a complete loop clockwise (the only railway spiral in Britain); it then passes over the original line just south of the station before turning north again to pass through a new tunnel at Moelwyn—opened in 1977—and thence along the west side of the Tan-y-Grisiau reservoir.

The line was reopened throughout its length to a new terminus in Blaenau Ffestiniog in May 1982.

1. BOYD J. I. C. *The Festiniog Railway.* Odhams Press, London, 1962.

2. LEWIS M. S. T. *How Festiniog got its railway.* Railway and Canal Historical Society, Caterham, 1965.

3. Tan-y-Bwlch Cast Iron Bridge (HEW 257) SH 647 415.

4. Funds galvanised Festiniog. *New Civil Engr*, 1984, 9 Feb., 36–37.

23. Ffestiniog Pumped Storage Scheme

HEW 1238
SH 679 445

This was the first pumped storage power station in Britain.[1] The upper reservoir was formed by enlarging Llyn Stwlan by a concrete dam 1250 ft long and 110 ft high; the lower, 1033 ft below, was created by damming the Afon Ystradau near the village of Tan-y-Grisiau. Here the concrete dam is 1855 ft long and 50 ft high. From the upper reservoir two vertical shafts fall 640 ft to two pressure tunnels which slope towards the power station for 3750 ft and then feed four steel pipes leading to the turbines and pumps a further 700 ft away.

The four alternators, each of 90 MW output when driven by the turbines, can operate as motors to drive the pumps. The latter are uncoupled when the station is generating. For pumping, the pumps are started by the turbines and when synchronization of the frequency with the system has been reached, the alternators become driving motors and the water to the turbines is cut off.

The power station building on the west side of the Tan-y-Grisiau reservoir is of steel-framed construction faced in local stone. At 273 ft long, 72 ft wide and 66 ft high above ground, it is probably the largest stone building to be constructed in Wales since Harlech and Criccieth castles.

As at Dinorwig, care was taken with landscaping. The spoil from Stwlan dam was placed in the reservoir and the face of Tan-y-Grisiau dam is concealed from view by rock from the excavations. General design was by the former Central Electricity Generating Board with Freeman Fox and Partners, in association with James Williamson and Partners and Kennedy and Donkin, as consultants. The main contractors for the civil engineering work were the Cementation Co. Ltd. and Sir Alfred

McAlpine and Son Ltd. Work began early in 1957 and was completed in March 1963.

1. ROSEVEARE J. C. A. Ffestiniog pumped storage scheme. *Proc. Instn Civ. Engrs*, 1964, **28**, May, 1–30.

24. Llechwedd Slate Caverns

HEW 1169
SH 698 475

The Ordovician slate at Llechwedd lies in five beds inclined at 30° to the horizontal and is sandwiched between layers of hard chert. It was discovered in 1849 by John W. Greaves. It was mined by boring and shot-firing using gunpowder, to form alternating chambers and pillars, which were carefully aligned vertically through over 1000 ft of rock to ensure support for all 16 levels in the mine.

There are five levels above the present-day tourist level, and ten below. True mining is only carried out on the second level below the tourist level, modern quarrying techniques being used from the surface to remove the pillars in the uppermost four former levels. Dewatering used to be by surface water power, later replaced by electricity, but nowadays the mines are allowed to drain through an adit at a lower level. There is a museum containing some interesting slate processing machinery devised by Greaves and originally powered from an overhead shaft driven by an overshot waterwheel. The slate was taken to the dock at Porthmadog on the Ffestiniog Railway, to connect with which Greaves installed a railway incline at Llechwedd in 1854.

25. The Cob, Porthmadog

HEW 1192
SH 572 384 to
SH 584 378

The construction of the embankment known locally as Y Cob was sponsored by William Alexander Madocks, MP for Boston, to reclaim 7000 acres of land from the tidal waters of Afon Glaslyn about five miles before the river enters Tremadoc Bay. The rockfill embankment, 90 ft wide at its base, 18 ft wide at the top and 21 ft high, was built on rush matting over a length of approximately 1400 yd, between Boston Lodge—on a hilly peninsula called Penrhyn-isaf—at its eastern end, and the harbour of Porthmadog at its western end. It was built by John

The Cob, Porthmadog, with Boston Lodge Works, Ffestiniog Railway in the foreground

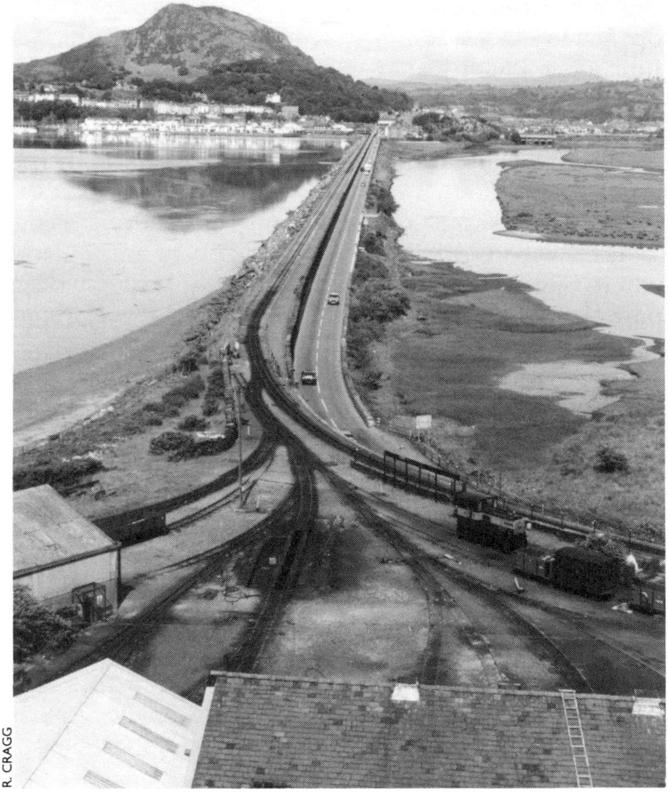

R. CRAGG

Williams between 1808 and 1811 and carries what is now the A487 road between Porthmadog and Maentwrog, the toll from motorists being collected at the eastern end.

The Ffestiniog Railway also makes use of the embankment on the southern side of the road.

26. Cambrian Coast Railway Lines

HEW 1227
SH 375 350 to SN 697 980 and SN 585 816

The Cambrian Coast lines round Cardigan Bay connect Pwllheli, Criccieth, Porthmadog, Harlech, Barmouth, Tywyn and Abersytwyth to the rest of the railway network via Dovey Junction and Machynlleth. They also form a standard gauge link between the now active narrow gauge lines—the Great Little Trains of Wales—such as the Ffestiniog, Talyllyn and Vale of Rheidol.

Authorized in 1861–62, the Cambrian lines were opened in 1863–67. They extend for 55 miles from Dovey Junction to Pwllheli, and about 15 miles to Aberystwyth. The railway skirts the sea briefly at Pwllheli, Afon Wen, Criccieth and Harlech; then from Llanaber, north of Barmouth, to Tywyn, almost continuously. Further stretches occur on both sides of the Dovey estuary, making a total of about 30 miles which is subject to coastal hazards.

The most difficult sections to maintain are at Llanaber, where huge concrete blocks connected by chains form part of the railway protective works, and at Friog near Fairbourne, where the line is carried on a cliff shelf and protected from rockfalls by an avalanche shelter. This has a 12 in. reinforced concrete roof slab some 60 yd long supported on reinforced concrete arches.

Crossing the Mawddach estuary, south of Barmouth, is a remarkable viaduct.[1,2] A rock foundation exists only at the northern end where there are two fixed spans of 37 ft 9 in. The navigable channel, which is near the north shore, was originally crossed by an unusual opening span which tilted and drew back over the track. Now there is a swing span of 136 ft with a central pivot and a fixed span of 118 ft, both hogback trusses carried on cylindrical piers which replaced the original cast-iron screw piles. To the south of the channel there are 113 openings of 18 ft span on timber pile trestles, which have either been replaced, or encased in glass fibre reinforced concrete sleeves, as a protection against attack by marine borer.

1. Barmouth Viaduct (HEW 1167) SH 623 151.

2. CONYBEARE H. Description of viaducts across the estuaries on the line of the Cambrian Railway. *Min. Proc. Instn Civ. Engrs*, 1870–71, **32**, 137–145.

27. Talyllyn Railway

HEW 1210
SH 586 005 to
SH 671 064

In the 1830s, slate began to be quarried at Bryn Eglwys, a remote site above the village of Abergynolwyn about 7½ miles north-east of Tywyn on the Welsh coast.

D. MITCHELL/TALYLLYN RAILWAY

Dolgoch Viaduct, Talyllyn Railway

Packhorses took the finished slates to Aberdovey, but following the purchase of the quarries by William McConnel, a railway line was planned from Bryn Eglwys to Aberdovey. After the opening of the Cambrian and Welsh Coast Railway between Aberdovey and Llwyngwril in 1863, the terminus of the line was changed to Tywyn to enable traffic to be interchanged with the new coastal railway.

The Engineer of the line was James Swinton Spooner, whose father and brothers were associated with the construction and development of the Ffestiniog Railway. Built to a gauge of 2 ft 3 in. (the minimum allowed by the Act), the line runs almost straight from Tywyn (Wharf) station, where there is a narrow gauge museum, to Abergynolwyn station, a distance of just over 6½ miles, climbing almost continuously along the south side of the valley of the Afon Fathew. Public passenger services terminated at Abergynolwyn but the line continued for a further ¾ mile to the foot of the first of several inclines connecting with the quarry workings. The maximum gradient of the line is 1 in 60. There were several intermediate stations, the most important being at Tywyn (Pendre), where the workshops and engine sheds were situated and where the passenger services from Abergynolwyn terminated. Just west of Dolgoch station, the line crosses the Nant Dolgoch at SH 650 045 on a three span brick viaduct about 50 ft above the stream, the line's most impressive feature. The railway was opened in December 1866.

By 1911 the controlling interest in the railway had passed from the McConnel family to Henry (later Sir Henry) Haydn Jones MP. The slate quarries finally closed in 1946 and after the death of the owner in 1950 the future of the line was in jeopardy, but a Preservation Society was formed and took over the running of the line, which now forms a major tourist attraction.[1] For a time L. T. C. Rolt, the author of several books on engineering history, acted as general manager. In 1976 an extension of the public passenger service beyond Abergynolwyn to a new terminus at Nant Gwernol was opened.

Recently, locomotive No. 7 has been named *Tom Rolt* in recognition of his contribution to the rescue of the Talyllyn Railway.

1. ROLT L. T. C. *Railway adventure*. David and Charles, Newton Abbot, 1970.

28. Llanrwst Bridge

HEW 164
SH 798 615

This well-proportioned three-arch masonry bridge carries a road to Gwydir Castle over the Afon Conwy in a most attractive setting. It is known as Pont Fawr, the 'Great Bridge' or 'Shaking Bridge', as it vibrates if the parapet is struck at a point above the central arch, which has a span of 60 ft. The side spans are 45 ft and the bridge was built in 1636, reputedly by Inigo Jones. The bridge has a total length of 169 ft and the width between the masonry parapets is 13 ft 2 in. They carry on their outsides, at the centre, coats of arms and the date of construction.

Centrally placed on top of one parapet is a bronze sundial erected to mark the tercentenary of the bridge, unfortunately at a point where it is inadvisable for pedestrians to linger because of the traffic and the narrowness of the bridge. The construction of the segmental arch rings is unusual. Although the lower voussoir rings are

R. CRAGG

Llanrwst Bridge

of normal shape with a slight taper inwards, the 4 in. by 20 in. slabs forming the second ring are dressed and laid to the curve with their long concave faces downwards on top of the lower voussoir ring.

29. Pont Carrog

HEW 682
SJ 115 437

On the upper Dee is Pont Carrog which was probably built in the seventeenth century. It is constructed of local stone and carries the B5436 Llidiart y Parc to Carrog road. It has five segmental arches and the cutwaters extend up to the parapet walls to form recesses at road level. Its total length is 186 ft and the roadway extends the full width of 12 ft between the parapet walls.

Upstream of Pont Carrog are three more interesting bridges over the River Dee.[1] The 1704 bridge at Corwen (SJ 069 434) is the longest bridge of the upper Dee, with seven spans and a total length of 107 yd. Near Cynwyd is Pont Dyfrydwy (SJ 052 412) with four segmental masonry arches and dating from about 1700. Pont Cilan (SJ 021 374) near Llandrillo was constructed at about the same period and has two masonry arches.

Pont Carrog

1. JERVOISE E. *The ancient bridges of Wales and Western England*. Architectural Press, 1936; republished by E.P. Publishing, East Ardsley, Wakefield, 1976, 14–16.

30. Vyrnwy Dam And Vyrnwy Aqueduct

HEW 214
SJ 018 193

The building of Vyrnwy dam to supply water to the City of Liverpool marked the introduction to Britain of the high masonry dam. Work began in July 1881; water from Lake Vyrnwy first reached Liverpool in 1891; and the works were formally opened by the Duke of Connaught in July 1892.[1,2]

The ground conditions were ideal for the construction of the gravity dam, which is 1175 ft long and has a maximum height of 145 ft from the foundation to the crest of the spillway section. At this point the dam is about 127 ft wide across the base. This was the first high dam designed to act as a weir and so dispense with a separate spillway.

The Engineers, Deacon and Hawksley, adhered to the two fundamental principles of great weight and watertightness, and much care was taken to achieve these

Vyrnwy Dam

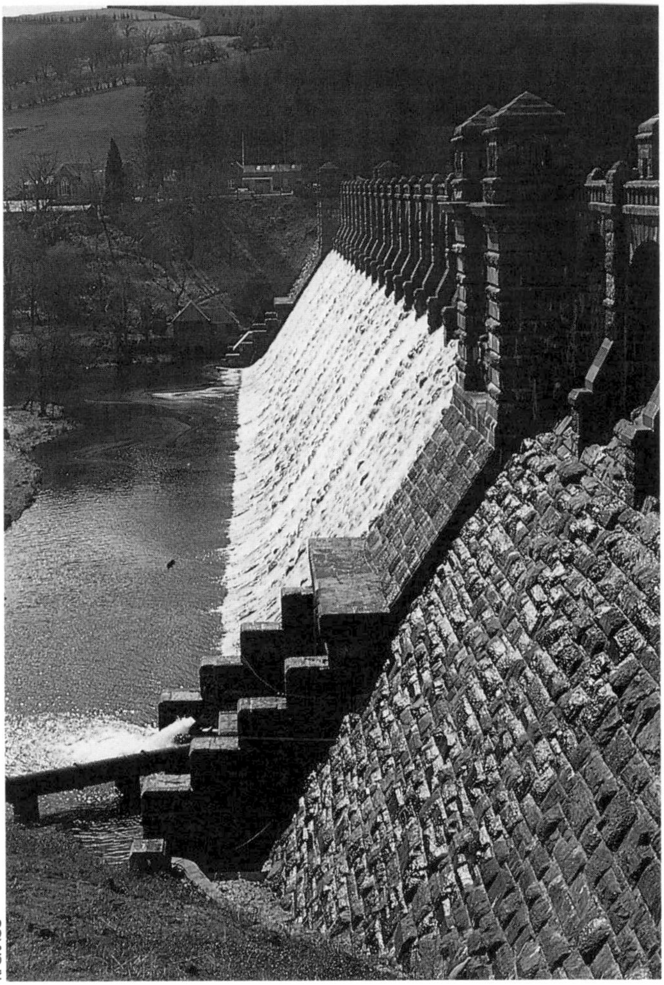

R. CRAGG

objectives. The mass of the dam consists of large irregularly shaped stone blocks, up to 10 tons in weight, set close together bedded on cement mortar. The spaces between them were then filled with mortar into which smaller broken stones were forced.

An important feature of the design was the provision of a drainage system in the foundations to prevent any build-up of water pressure on the underside of the base, possibly leading to overturning.

The aqueduct is 68 miles long and there are now three

HEW 1147
SJ 01 20 to
SJ 47 93

42 in. diameter pipes to deliver up to 50 million gallons per day to Prescot service reservoirs east of Liverpool. The route follows the Dee–Severn watershed to maintain high ground until the basins of the Mersey and Weaver are reached.

The first two pipelines were generally of cast iron but the use of riveted steel pipes to facilitate maintenance in the 9 ft diameter cast-iron tunnel under the Mersey marked an early use of steel for trunk water mains.

The first three tunnels, at Hirnant, Cynynion and Llanforda, are alike, with brick and concrete linings to protect against leakage. Hirnant was later duplicated at Aber to enable maintenance to be carried out.

At the Oswestry reservoir, where the water is filtered, a 500 yd long earth dam impounds some 52 million gallons storage. Two booster pumps at Oswestry were refurbished in 1991.

There are several balancing reservoirs and water tanks. The largest tank, at Malpas, has a capacity of 4.5 million gallons, while the Norton tank, of 650 000 gallons capacity, is housed within a monumental sandstone tower some 110 ft high.[3,4]

The first section of the third pipe was laid in 1926–38 in steel. This saw the beginning of the more general use of bituminous coated steel pipes for trunk water mains in place of cast iron, which had been in use since 1810.[5]

After 1946, to increase capacity, a fourth pipeline was laid upstream of Oswestry, with three booster stations downstream thereof.[6] The pipe crossings under the Mersey and the Manchester Ship Canal were rearranged in 1978–81.

1. Deacon G. F. The Vyrnwy works for the water supply of Liverpool. *Min. Proc. Instn Civ. Engrs*, 1895–96, **126**, 24–67.

2. Binnie G. M. *Early Victorian water engineers*. Thomas Telford, London, 1981, 143–148.

3. Deacon G. F. The Vyrnwy works and the Vyrnwy valley water supply to Liverpool. *Engineer*, London, 1892, **73**, 15 July, supplement, 13–15.

4. Norton Water Tower (HEW 1148) SJ 554 817.

5. Stilgoe J. H. T. Trunk mains for water supply. *Trans. L'pool Engng Soc.*, 1949, **72**, Feb., 117–151.

6. White W. F. Vyrnwy aqueduct, fourth instalment pipeline. *J. Instn Water Engrs*, 1950, **4**, Feb., 13–67.

R. CRAGG

Llangollen Ancient Bridge

31. Llangollen Ancient Bridge

HEW 165
SJ 215 422

This example of early road engineering is sometimes described as one of the Three Jewels of Wales.

The bridge, built on this site in 1282, was reconstructed in sandstone about 1500 to the present style with four arches, three of which are pointed and the fourth segmental. It has piers with cutwaters that extend up to the parapet walls forming recesses. In 1865 it was extended across the railway by an additional span of iron girders and dressed freestone.

The bridge was widened on the upstream side from its original 12 ft in 1873 to 20 ft between the parapets and again in 1969 to 36 ft.

32. Shropshire Union Canal, Llangollen Branch

HEW 1204
SJ 196 433 to
SJ 370 318

The route and history of the Ellesmere Canal (later part of the Shropshire Union) is dealt with in more detail in Chapters 7 and 8. Briefly, the section of the original main line projected to run between the rivers Severn and Dee, through the hilly Welsh border area, was never completed.

By the time the decision had been made to change the originally projected north–south route of the Ellesmere Canal to the more easterly route, progress had been made up the Dee valley towards Llangollen, including the building of the Pontcysyllte Aqueduct to take the canal across the river at a high level. This section of canal, from Frankton to Trevor, therefore became a long branch and the originally proposed branch, from Frankton to the Chester Canal at Hurleston via Whitchurch, became the main line. The branch was completed in 1805 under an Act of 1804 with an extension to Llantysilio, just to the west of Llangollen. This was done to obtain an adequate water supply, which is fed along the main line of the canal to the reservoir at Hurleston (SJ 626 553) at the junction with the Chester Canal, a fall of some 123 ft. The Chester Canal had drawn its supply from a feeder above Bunbury Top Lock, but this was scarcely sufficient.

It is interesting to speculate that if the change of route of the Ellesmere Canal had been decided upon earlier there might have been no need to build the magnificent Pontcysyllte Aqueduct, as the water supply could have been taken down the right bank of the Dee instead of the left.

33. Horseshoe Falls Weir, Llantysilio

HEW 1235
SJ 196 433

The Horseshoe Falls Weir was built by Thomas Telford to secure a water supply from the River Dee at Llantysilio to feed the Ellesmere Canal. It is not clear whether the name derives from the shape of the weir, which is actually J-shaped, from the horseshoe bend in the river, or even from the corresponding bend in Telford's Holyhead Road above Llangollen. From this road there is an excellent view of the weir and its picturesque surroundings.

The weir is sited above rapids which are now an international venue for canoeing events. It is of masonry, 460 ft long with an upstream slope and a vertical downstream face 4 ft high. The crest is of 4 ft square stones with a flat bull-nosed cast-iron capping in 9 ft lengths, each

R. CRAGG

Horseshoe Falls

length secured to the masonry by three lugs. A spare length of capping is stored on site.

In addition to the construction of the Horseshoe Weir, in 1808 Telford obtained permission from the local landowner to raise the level of Llyn Tegid (Bala Lake) 2 ft 6 in. by means of a weir with sluices, to ensure an adequate reservoir. Similar arrangements were made at Llyn Arenig Fach and Llyn Arenig Fawr.

Much of the water from the River Dee now goes to water treatment plants for public supplies as well as for canal impounding; and in more recent years the security of supply for Deeside and Merseyside has been augmented by the construction of reservoirs at Llyn Celyn (SH 936 356) in 1965 and Llyn Brenig (SH 822 426) in 1976.

34. Pontcysyllte Aqueduct

HEW 112
SJ 271 420

Pontcysyllte Aqueduct carries the Llangollen Branch of the Ellesmere Canal over the River Dee 2 miles west of Ruabon and was commenced in 1795 and completed in 1805. It was scheduled as an ancient monument in 1958.

The project attracted great admiration throughout the whole country. Robert Southey wrote of 'Telford who o'er the Vale of Cambrian Dee aloft in Air at giddy height upborne carried his navigable road', whilst Sir Walter Scott described it as the greatest work of art he had ever seen.

Before Telford's pioneering cast-iron aqueduct was built on the Shrewsbury Canal at Longdon on Tern (see Chapter 7), canal aqueducts had generally consisted of a heavy puddled clay waterway constructed on squat brick arches. Telford's scheme for the crossing of the River Dee developed further the use of cast iron. At Pontcysyllte, a trough 11 ft 9¾ in. wide is carried on four arch ribs over each of the nineteen 44 ft 6 in. spans between the masonry piers. The aqueduct is 1007 ft long and at the highest point is 121 ft above the river.

The towpath, which has a protective iron parapet railing, overhangs the channel on the east side, thus giving space for the movement of water displaced by the passage of boats, whilst leaving a clear width of 7 ft 10 in. This is in contrast with Telford's earlier iron aqueduct at Longdon on Tern where the towpath is carried alongside the trough, level with the trough base. The slender masonry piers, which are partly hollow, taper upwards to 13 ft by 7 ft 6 in. at the top. There is an interesting

Pontcysyllte Aqueduct

R. CRAGG

commemorative tablet fixed to the base of the pier adjoining the south bank of the river.

The embankment at the south end was one of the greatest earthworks undertaken at the time.

The aqueduct is important historically in that it brought together Thomas Telford, the designer of the aqueduct, William Jessop,[1] the Engineer to the Ellesmere Canal Company (who approved the design), Mathew Davidson, who was Telford's supervising engineer, William Hazledine, the local iron master, and the two master masons John Simpson and John Wilson. Members of this team were subsequently engaged on many famous civil engineering works, ranging from the Caledonian Canal to the Menai Bridge.

Many pictures of the aqueduct show in the foreground the Cysyllte Ancient Bridge,[2] a three-span sandstone arched bridge built in 1696 across a bend in the river.

1. HADFIELD C. and SKEMPTON A. W. *William Jessop, engineer*. David and Charles, Newton Abbot, 1979, 222–228.

2. Cysyllte Ancient Bridge (HEW 160) SJ 268 420.

35. Chirk Aqueduct

HEW 111
SJ 286 371

This major work carries the Llangollen Branch of the Ellesmere Canal some 70 ft above the valley of the River Ceiriog. It was built between 1796 and 1801.

The structure appears from the exterior to consist of masonry piers and arches giving clear spans of 40 ft and is 710 ft long overall. However, Telford used cast-iron plates to form the bed of the 5 ft deep channel, thus enabling him to reduce the depth of the masonry, and hence the weight, above the piers. The plated bed was flanged at the edges and secured by nuts and bolts at each joint and, in addition, was built into the masonry at each side. Side plates were added circa 1870.

The sides of the waterway were waterproofed by ashlar masonry and hard burnt bricks in Parker's cement, thus obviating the need for clay puddle used by the earlier canal engineers.

The adjacent 16-span railway viaduct[1] was built in 1848 originally in laminated timber by Brassey, Mackenzie and Stephenson with Mr Meakin as Agent for the

Shrewsbury and Chester Railway. It was rebuilt in brick and stone in 1859.

1. Chirk Railway Viaduct (HEW 602) SJ 286 372.

Chirk Aqueduct and Railway Viaduct

36. Chirk Canal Tunnel

HEW 162
SJ 285 374 to SJ 284 378

Chirk Tunnel, at the north end of Chirk Aqueduct, was built by Thomas Telford in 1801. At 459 yd long, it is the longest of three tunnels on the Llangollen Branch of the Ellesmere Canal. All three tunnels were unusual for the time in having a towpath taken through the bore, thus obviating the need for 'legging' the boats through the tunnel. The portals make the tunnel appear larger than it is, as the bore is flared at the ends.

37. Cefn Viaduct

HEW 568
SJ 285 412

This handsome viaduct, built by Brassey, Mackenzie and Stephenson to the designs of Henry Robertson in 1846–48, carries the double line of the former Shrewsbury and Chester Railway (later the Great Western Railway, although it never became broad gauge) across the River

R. CRAGG

R. CRAGG

Dee and adjacent fields about a mile downstream from the Pontcysyllte Aqueduct. The abutments and piers are of stone and the arches are of brick with stone facings. There are 19 openings of 60 ft span and two of 15 ft. The total length is 510 yd and the greatest height 148 ft. It was claimed to be the longest viaduct in Britain at the time of its building.

38. River Dee Viaduct

HEW 1782
SJ 300 410

Claimed to be the highest road over river bridge in Britain, this concrete box girder structure was erected in 1987–90 to carry the Newbridge bypass (A483) over the valley of the River Dee not far from Chirk. The viaduct has five spans, the central span being 83 m, and the deck is 12.3 m wide. The four reinforced concrete piers are all hollow and up to 55 m in height; they support a post-tensioned prestressed concrete deck constructed on the balanced cantilever principle using variable depth single-cell box girders.

The viaduct was designed by Travers Morgan and Partners and constructed by Edmund Nuttall Ltd.

39. Welshpool and Llanfair Light Railway

HEW 1483
SJ 229 072 to
SJ 106 068

This narrow gauge railway, running through the magnificent scenery of the Welsh border country between Welshpool and Llanfair Caereinion, was opened on 4 April 1903 and was built under the provisions of the Light Railways Act 1896.[1]

Built to a gauge of 2 ft 6 in., the railway started in the yard of the Cambrian Railway's station at Welshpool where there were interchange sidings and locomotive and goods sheds. Passing through the streets, it emerged on the west side of the town at Raven Square, the site of the present Welshpool terminus of the line. From Raven Square the railway faces a steep climb, including a 1 mile section at 1 in 29 (3.5 per cent) to reach its highest point at Golfa Summit, 363 ft above the starting point. From there the route descends into the valley of the River Banwy, which it follows until it reaches Llanfair

Opposite: Chirk Canal Tunnel (above); Cefn Viaduct (below)

R. CRAGG

Welshpool and Llanfair Light Railway: Locomotive No. 14 at Raven Square Station

Caereinion station, 9 miles from Welshpool. There are intermediate stations at Sylfaen, Castle Caereinion, Cyfronydd and Heniarth. As a true light railway, the line is steeply graded and sharply curved, the reverse curves at Dolrhyd Mill being of only three chains radius. There are few major structures on the line, two notable ones being the six masonry arch Brynelin Viaduct and a substantial three-span steel girder bridge which carries the line over the River Banwy. The Engineer for the building of the railway was A. J. Collin, then Chief Engineer of the Cambrian Railway, and the contractor was John Strachan.

From its opening the line was worked by the Cambrian Railway and both passengers and freight were carried. Locomotive power was provided by two identical 0–6–0 tank engines supplied by Beyer Peacock in 1902 and named *The Earl* and *The Countess* in honour of the Earl and Countess of Powis, owners of nearby Powis Castle.

In 1923 the railway passed into the hands of the Great Western Railway and in 1931, in the face of increasing competition from road vehicles, the passenger service was discontinued. A freight service continued to be operated, the principal traffic being coal, agricultural implements, fertilizers and building materials westwards to Llanfair and timber, livestock and other agricultural produce eastward to Welshpool. In 1956 the railway was closed to all traffic, the service in the final years having been reduced to one train a day on most weekdays.

A Preservation Society was formed in 1956 and in 1960 the Welshpool and Llanfair Light Railway Preservation Company was incorporated. Based at a new headquarters at Llanfair Caereinion, the new company restarted passenger services in 1963, at first running only to Castle Caereinion but later extended to Sylfaen station. Not until 1981 were services finally extended once again to Welshpool, not to the original terminus, for the last 1 mile of the town section had been lifted, but to a new terminal station at Raven Square. With the two original locomotives and an interesting collection of motive power and rolling stock from around the world, volunteer train crews continue to provide a fascinating glimpse of steam-hauled, light railway operation for the numerous tourists and enthusiasts who visit this part of the Welsh borderland.

1. Cartwright R. and Russell R. T. *The Welshpool and Llanfair Light Railway*. David and Charles, Newton Abbot, 1989.

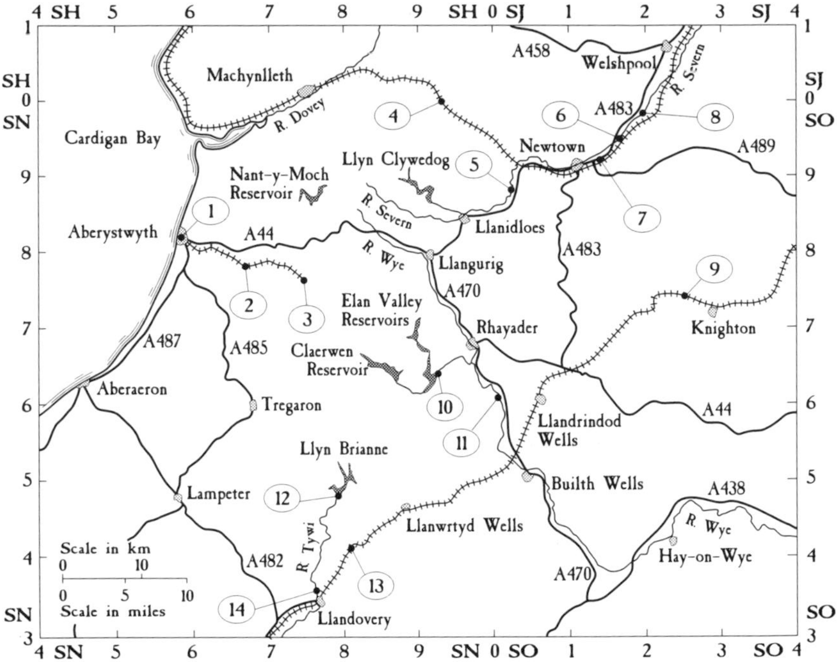

1 Aberystwyth Cliff Railway
2 The Vale of Rheidol Railway
3 Devil's Bridge
4 Talerddig Cutting
5 Llandinam Bridge
6 Abermule Bridge, Brynderwen
7 Penarth Weir, Newtown
8 Caerhowel Bridge
9 Cnwclas Viaduct
10 Caban Coch Dam
11 Doldowlod Bridge
12 Llyn Brianne Dam
13 Cynghordy Viaduct
14 Dolauhirion Bridge

2. Mid-Wales

The area covered by this chapter extends eastward from the shores of Cardigan Bay to the English border and is bounded to the north by the River Dyfi (Dovey) and to the south by the mountain ranges beyond which lie the valleys of South Wales. It is predominantly a land of mountains, forests and rivers. Both the Wye and the Severn have their birthplace on the slopes of Plynlimon and flow south-east and north-east respectively through the region.

The landscape and high rainfall of the area have attracted the attentions of water supply engineers and man-made reservoirs abound. Of particular note is the Elan valley group to the west of Rhayader, which supplies Birmingham via a 73 mile long aqueduct. The waters of the Tywi and the Clywedog have also been impounded for this purpose but in these more modern schemes the rivers themselves are used as the means for carrying the water to the areas where it is required.

Sheep and cattle farming are the major industries but there are both inland and seaside resorts, some served by the remaining rail network, while traces of ancient industries lie near to the occasional modern factory. Most of the works described in this chapter are associated with the provision of transport facilities in an area of rough topography and fast-flowing rivers.

1. Aberystwyth Cliff Railway

HEW 1130
SN 583 826

The Cliff Railway at the northern end of Aberystwyth promenade was built for the Aberystwyth Improvement Company and opened on 1 August 1896. It is a funicular railway. The two passenger cars run on parallel 4 ft 8½ in. gauge rail tracks and each is attached to a continuous wire rope cable which is operated by a stationary electric motor. The cars each have a capacity of about 30 persons. Their self weights are mutually balancing, one car being raised as the other one is lowered. Until 1921 the railway was operated by a water balance system.

An early photograph indicates that substantial earthworks were involved in achieving an approximately uniform gradient for the track, which rises nearly 400 ft over its length of 798 ft. The railway was designed by G. Croydon Marks, later Baron Marks of Woolwich.

O. M. GIBBS

Car ascending from the lower station, Aberystwyth Cliff Railway

2. Vale of Rheidol Railway

HEW 1121
SN 585 816 to
SN 739 770

The Vale of Rheidol Railway, 11¾ miles long, is a 1 ft 11½ in. gauge line, built to transport lead and zinc ore from the area around Devil's Bridge to the harbour at Aberystwyth. The Engineers for the line were Sir James and William Szlumper and the contractors were Messrs Pethick Brothers. The line opened to goods traffic in August 1902 and to passengers in the following December.[1]

It is now solely a passenger line and operates during the spring and summer months, when it carries many thousands of tourists through the scenic beauties of the Rheidol valley. Haulage is normally by steam traction and the three steam locomotives are named *Prince of Wales*, *Owain Glyndwr* and *Llewelyn*; there is also a diesel hydraulic locomotive. Ownership of the line passed to the Cambrian Railway in 1913, and to the Great Western Railway in 1922. It became part of British Railways Western Region in 1948 and, following the demise of steam traction in the 1960s, was British Rail's only steam operated line. In 1989 it became privately owned.

1. BOYD J. I. C. *Narrow gauge railways in mid-Wales (1850–1970).* Oakwood Press, Oxford, 1986, 192–228.

3. Devil's Bridge

HEW 1120
SN 742 771

Situated near Devil's Bridge station, the upper terminus of the Vale of Rheidol Railway, Devil's Bridge is a popular mid-Wales tourist attraction. Three bridges cross the Afon Mynach, one above the other. The first and lowest is a medieval pointed arch in stone of 15 ft span, traditionally thought to have been built by the monks of Strata Florida Abbey. The second bridge is considered to date from the mid-eighteenth century and is a flat segmental arch of 32 ft span.

Early in the nineteenth century the spandrel walls surmounting this arch were raised to allow the approaches to be less steep and at the beginning of the twentieth century a steel bridge was built some 7 ft above the earlier structure. This bridge, since repaired and strengthened, continues to carry traffic to the present day.

Access into the deep gorge is through turnstiles via

Devil's Bridge

R. CRAGG

steps and pathways from which the bridges and the nearby waterfalls can be seen. Motorists must use car parks in the vicinity.

4. Talerddig Cutting

HEW 1829
SN 931 995 to
SH 930 001

The Newtown and Machynlleth Railway was built under an Act of 1856. Its route between the two towns of its name leaves the valley of the River Severn at Caersws and climbs over the hills to descend into the valley of the River Dovey at Cemmaes Road.

At Talerddig the single line reaches its summit level of 693 feet above sea level. Here the original plan was for the line to tunnel through a steep rock outcrop but the plans were changed and a deep cutting was excavated. Talerddig cutting is about 120 feet deep at its maximum and is one of the deepest on British railways. The stone excavated from the cutting was used in the construction of many of the masonry structures on the line.

The Engineers for the railway were Benjamin and Robert Piercy and the contractors were David Davies and Thomas Savin. It was opened in January 1863.

R. CRAGG

Talerddig Cutting

R. CRAGG

Llandinam Bridge

5. Llandinam Bridge

HEW 850
SO 025 886

This cast-iron arch road bridge over the River Severn was built in 1846. The single arch spans 90 ft with a rise of 9 ft and is made up of three ribs, each consisting of five segments; each segment in turn has five X lattice panels. The spandrels are also of open X pattern with 4 in. cruciform members, generally similar to those of the Mythe Bridge at Tewkesbury. The bridge is stiffened laterally by rectangular and circular cross-members connecting the arch ribs. It has a width of 10 ft 5 in. and carries a minor road.

It was the first cast-iron bridge to be built in the county of Montgomery and was cast by the Hawarden ironworks to the requirements of Thomas Penson, County Surveyor, who also designed the bridges at Abermule and Caerhowel. It currently carries a 3 ton weight limit.

At the east end of the bridge, adjacent to the main road, there is a statue of David Davies, the eminent railway contractor, owner of the Ocean Coal Company and the builder of Barry Docks, who was born in Llandinam in 1818. He died in 1890.

6. Abermule Bridge, Brynderwen

HEW 342
SO 162 951

The elegant bridge over the River Severn just to the north-east of Abermule was cast by the Brymbo Company iron foundry in 1852. The single 110 ft span arch, with a rise of 12 ft 3 in., consists of five 2 ft 9 in. deep ribs, each with seven bolted segments. The abutments are of stone. An inscription cast into the outer two arch ribs records that it was the second iron bridge in the county of Montgomery. The width of 21 ft between the parapet rails includes a 17 ft carriageway.

Although still in use, the bridge is no longer on the main road as this now bypasses Abermule.

7. Penarth Weir, Newtown

HEW 1279
SO 140 927

In 1819 the final section of the Montgomeryshire Canal was opened from its temporary terminus at Garthmyl to Newtown.

In order to ensure an adequate supply of water to the upper end of the canal at Newtown, the Engineer, Josias Jessop, and his Resident Engineer, John Williams, constructed a weir across the River Severn below Newtown,

Penarth Weir

R. CRAGG

from which a short feeder channel conveyed water diverted from the river into the canal.[1]

The masonry weir has a total height of 8 ft, in two steps. The upper crest is 3 ft wide, followed by a slope about 10 ft long leading to an apron 11 ft 10 in. wide; this slopes up to the lower crest, which is 3 ft 6 in. wide, with a final slope down to the base of the weir. The weir is curved in plan, forming a segment of a circle of about 150 ft radius. It is convex upstream and the crest length is about 140 ft.[2]

On the south bank of the river the weir abuts a vertical rock face. On the north side there is a bypass sluice and a fish ladder, both of which are of relatively modern construction. Just above the weir is the entrance to the canal feeder, which runs north-eastward parallel to the river for about 350 yards to join the canal.

1. HADFIELD C. *The canals of the West Midlands*. David and Charles, Newton Abbot, 1985, 189–196.

2. BINNIE G. M. *Early dam builders in Britain*. Thomas Telford, London, 1987, 54–56.

8. Caerhowel Bridge

HEW 851
SO 197 982

The third in the series of cast-iron bridges over the River Severn is situated near Caerhowel Hall, some 2 3/4 miles downstream from Abermule Bridge. It was also cast by the Brymbo Company but rather later, in 1858. It has two spans of 72 ft 8 in., each with five ribs, and each rib has five segments.

The bridge is currently relieved from loading by a Bailey bridge and single-line traffic is controlled by traffic signals.

9. Cnwclas Viaduct

HEW 594
SO 250 742

In the late 1850s, several railways were proposed to link the lines of South Wales with the West Midlands. One of these was built as the Central Wales line running from Craven Arms to Llandovery.

This is in picturesque country. The Cnwclas Viaduct carries the single line of the railway over a side valley of the River Teme as the line starts its climb to the summit at Llangynllo Tunnel. The viaduct, designed by Henry

Cnwclas Viaduct

Robertson and built by Thomas Brassey, is an elegant structure. There are 13 arches of 30 ft span, the maximum height above the valley being 75 ft. Semicircular stone arches surmount stone piers and the whole viaduct is richly decorated with castellated stone towers at each end and a battlemented parapet.

The section of the Central Wales line upon which the viaduct stands was opened on 10 October 1863.

10. Caban Coch Dam

HEW 550
SN 925 646

In 1892 the Birmingham Corporation Water Act authorized the construction of reservoirs in the Elan and Claerwen valleys, south-west of Rhayader, and of an aqueduct to convey water to the city. James Mansergh was the Engineer.[1] The initial works in the Elan valley comprised three reservoirs which were built by direct labour and completed in 1904.

Caban Coch is the first dam up the valley from Rhayader. It is 610 ft long, 122 ft high and 5 ft wide at the crest and is built of cyclopean mass concrete faced with block-in-course masonry. The downstream face has an inwardly curved batter, struck to a 340 ft radius, to within

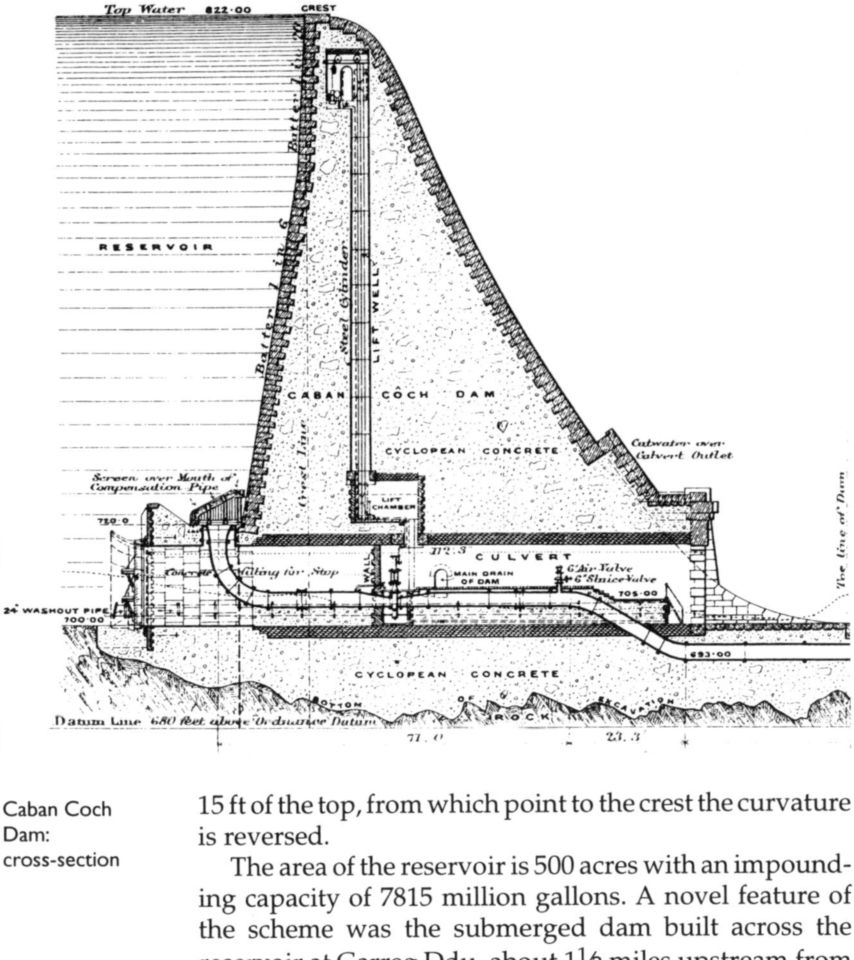

Caban Coch Dam: cross-section

15 ft of the top, from which point to the crest the curvature is reversed.

The area of the reservoir is 500 acres with an impounding capacity of 7815 million gallons. A novel feature of the scheme was the submerged dam built across the reservoir at Garreg Ddu, about 1½ miles upstream from Caban Coch. At times when the reservoir is very low this keeps the water at the required level to feed the aqueduct, while the water impounded in the reservoir below it can be used to provide compensation water.

1. MANSERGH E. L. and MANSERGH W. L. The works for the supply of water to the City of Birmingham from mid-Wales. *Min. Proc. Instn Civ. Engrs*, 1911–12, **190**, 23–25.

11. Doldowlod Bridge

HEW 1245
SO 003 617

This privately owned suspension footbridge over the River Wye about 5 miles south of Rhayader lies within the Doldowlod Estate of the Gibson-Watt family. The suspension chains span 120 ft between 14 ft 6 in. high cast-iron towers, each of which is built up from five separate castings, bolted together; each leg has an upper and a lower section with a decorative cross-member. The links of the chains consist of ¾ in. diameter iron rods 7 ft 4 in. long, the number of rods in each link varying from five at the towers to two at mid-span. The links are connected by 1 in. diameter bolts. Inclined hangers attached to the chains at each link connection support the timber deck, an arrangement similar to that used by Dredge for his suspension bridges such as that over the Kennet and Avon Canal at Stowell Park and Victoria Bridge, Bath.

The bridge was built about 1880, the exact date being uncertain. The ironwork was cast at Llanidloes Railway Foundry.

Doldowlod Bridge

R. CRAGG

12. Llyn Brianne Dam

HEW 552
SN 791 485

Llyn Brianne Dam and Reservoir lie in mountainous country to the west of Llanwrtyd Wells and north of Llandovery.

The dam, which was completed in 1972, stores and regulates the flow in the River Tywi in order that water may be abstracted at Nantgaredig some 40 miles downstream, to supply Swansea, Neath and Port Talbot.[1]

The crest is 900 ft long and 30 ft wide and the height of 300 ft makes it the highest dam in Britain. It is constructed of rock filling with a boulder clay core. There is car parking provision for visitors who may drive some 6 miles along the length of the reservoir, and viewing points from which may be observed the spectacular jet by which water is discharged from the foot of the dam.

The work was designed by Binnie and Partners and constructed by George Wimpey and Co. Ltd.

1. Carlyle W. J. and Owen R. C. The River Towy scheme for the West Glamorgan Water Board. *J. Instn Water Engrs*, 1973, **27**, 287–317.

Llyn Brianne Dam

R. CRAGG

13. Cynghordy Viaduct

HEW 272
SN 808 418

The Llandovery to Builth section of the Central Wales Railway line was opened in 1868. This section included a number of interesting engineering works, the viaduct across the Afon Bran being the most spectacular. The viaduct has 18 spans of 36 ft, is curved in plan and 93 ft high. The single-line railway is still in use.

The viaduct is about 1 mile from the village of Cynghordy, which lies on the road from Llandovery to Llanwrtyd Wells.

14. Dolauhirion Bridge

HEW 167
SN 762 361

William Edwards, his two sons and his grandson, built a number of bridges in Wales in addition to the world famous bridge at Pontypridd. Some of these, like Pontypridd, were single-span bridges with openings in the haunches, and the best example is that at Dolauhirion, built in 1773. It is still in daily use. The bridge spans the River Tywi approximately 1 mile from Llandovery and just off the road towards Rhandirmwyn and Llyn Brianne.

According to Jervoise[1] the Dolauhirion Bridge is the finest bridge over the upper part of the Tywi. He describes it as 'a single segmental arch with a span of 28 yards and the roadway 12 ft in width. The circular openings in the haunches of this bridge are a distinctive feature of Edwards' bridges.'

In the Dolauhirion Bridge only one opening was provided at each abutment, and the feature is not, of course, unique to Edwards.

1. JERVOISE E. *The ancient bridges of Wales and western England.* Architectural Press, 1936; republished by E.P. Publishing, East Ardsley, Wakefield, 1976, 73.

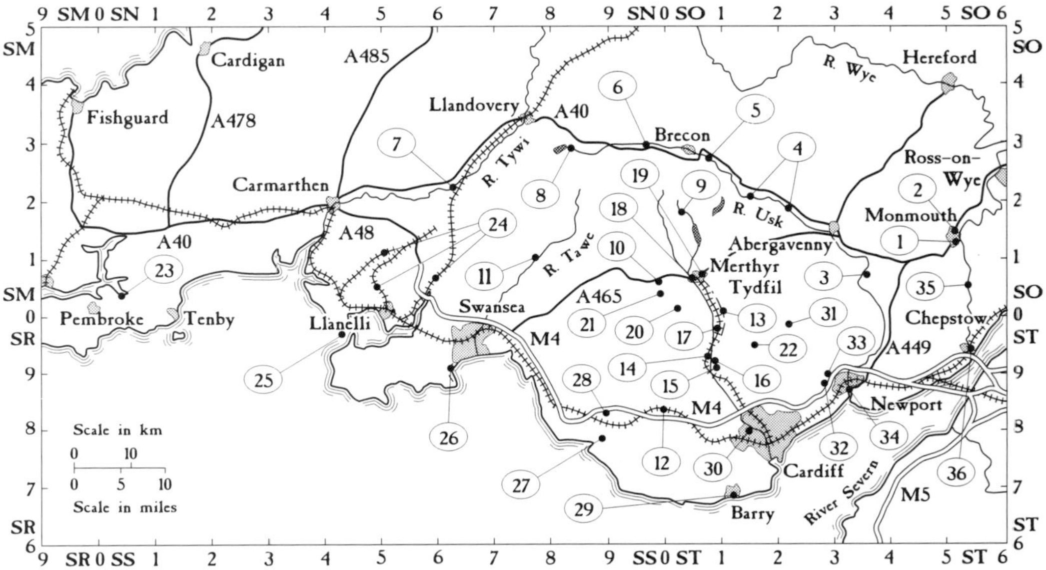

1 Monnow Bridge, Monmouth
2 Monmouth Forge Generating Station
3 Pant y Goitre Bridge
4 Crickhowell and Llangynidr Bridges
5 Brynich Aqueduct
6 Abercamlais Suspension Bridge
7 Llandeilo Bridge
8 Usk Dam
9 Upper Neuadd Dam
10 Nant Hir Reservoir
11 Twrch Aqueduct
12 South Wales Railway
13 Penydarren Tramroad
14 Taff Vale Railway
15 Pontypridd Bridge
16 Berw Road Bridge, Pontypridd
17 Pont y Gwaith
18 Cefn Coed y Cymmer Viaduct
19 Pont y Cafnau
20 Railway Bridge at Cwmbach, Aberdare
21 Robertstown Cast Iron Tram Bridge, Aberdare
22 Hengoed Viaduct
23 Carew Tide Mill
24 Three Railways near Llanelli
25 Whiteford Point Lighthouse
26 Swansea and Mumbles Railway
27 New Inn Bridge, Merthyr Mawr
28 Glanrhyd Bridge
29 Barry Docks
30 Melingriffith Water Pump
31 Crumlin Viaduct
32 Flight of Locks near Rogerstone, Monmouthshire Canal
33 Ynys-y-Fro Reservoir
34 Newport Transporter Bridge
35 Bigsweir Bridge
36 Chepstow Bridge

3. South Wales

The industrial history of South Wales has been determined by its geography and geology. Beneath most of the area lies a bowl-shaped coalfield, outcropping in the north near to the Brecon Beacons and in the south along a line a little to the north of the M4 motorway.

In early times the landscape was green woodland, but in the eighteenth and nineteenth centuries ironworks grew up along the northern and, to a lesser extent, the southern edges of the coalfield, to exploit the iron ore and limestones, harnessing the plentiful supplies of timber and water for power. Later the abundant coal resources were used to drive a massive expansion of industry in the area. Packhorse routes, canals and tramroads were established to take the products of the iron industry to the ports. From about 1840, as steam replaced sail, locomotives replaced horses and improved techniques were developed for working the deeper seams, so the output of coal, noted abroad as well as at home for its excellent steam raising qualities, grew enormously.

All this was not achieved without great effort, for the coalfield is crossed by a series of deep, narrow valleys running approximately north to south, in which construction is fraught with difficulties. The canals in the valleys which sloped steeply towards the sea had locks which were frequently deeper than was the practice in England, while the railways required numerous viaducts in timber, iron and masonry across the valleys, with tunnels through the intervening mountains.

The advent of oil firing and the internal combustion engine at the beginning of the twentieth century began a slow decline of the coal trade and its associated heavy industries until today only one deep mine remains. At present the major industrial complexes are concentrated along the seaboard, while the valleys strive to attract new industries. The main emphasis on communications has changed from north–south to east–west and the ports have been redeveloped to handle other cargoes in place of their coal exports and also as foci for the new industry of the area—tourism. Among the more recent industrial developments must be mentioned the extensive oil refinery at Milford Haven.

The coal-mining industry has preserved few remains of its past, as civil engineering techniques have been used in the clearance of the many hundreds of abandoned spoil tips in the process of major land reclamation operations.

As a result of these land reclamation operations some of the features of the valleys have disappeared, including Weaver's Mill at Swansea and the world famous Crumlin Viaduct. In their place, however, new engineering works have been constructed, with modern roads extending into the deep valleys. Port redevelopment has included the provision of major barrages as South Wales develops from an industrial society based on heavy industry into one centred around manufacturing and tourism, with the valleys often reverting to dormitory areas.

The South Wales Docks

From prehistoric times until the mid-nineteenth century, South Wales was more accessible by sea than by land. With mountainous country to the north, river estuaries were frequent and formed shelter, loading and unloading facilities for the sailing ships then in use. With the increasing volume of exports, chiefly iron and coal, and the problems posed by the large tidal range in the Severn estuary, the need for dock construction grew. More specifically the need was for 'wet' docks within which the water could be maintained at or near high tide level while shipping entered and left through a basin, isolated from both the dock and the sea by lock gates.

It was the last few hundred metres of the Glamorganshire Canal that formed, at Cardiff in 1798, the first effective dock. Streets named East and West Canal Wharf still survive from this era. The first wet dock at Cardiff, the Bute West Dock, was completed in 1839. It rapidly became inadequate and was supplemented by the much larger Bute East Dock, built in sections from 1855 to 1859. Construction of Cardiff docks was, however, preceded a few years by building at Burry Port and Llanelli. Fed by three of the early railways, described later in this chapter, their exports were anthracite and tinplate. At Swansea, trade included massive imports of copper ore. Brunel, succeeded by R. P. Brereton, engineered the Briton Ferry Docks which were linked to the South Wales Mineral Railway.

To the east, Newport Town Dock was opened in 1842; it was subsequently extended and eventually supplemented by the North and South docks, completed in 1914.

Similar development of a second phase of dock building took place at Cardiff (the Roath and Queen Alexandra Docks), at Penarth, Barry (considered in more detail later in this chapter) and Swansea (the South Dock,

Prince of Wales and Queen's Docks) but the coal trade reached its peak at the time of World War I and a long period of decline began.

Most of the docks, with the exception of Swansea, belonged to the railway companies which served them and, as a consequence, in 1923, following the grouping of the railways, the Great Western Railway became the largest owner of docks in the world.

Today, the most modern of the docks, at Newport, Cardiff and Swansea, remain in operation and in the ownership of Associated British Ports. The removal of the former extensive railway sidings and coal hoists has given space for the handling of bulk imports such as timber, for oil storage and for containerization. At the same time road access has been improved.

Finally, at Cardiff, a barrage is being built across the mouths of the Taff and Ely rivers. The 500 acre freshwater lake thus formed is expected to rejuvenate areas of derelict land and mud flats where once stood the busiest coal-exporting port in the world.

1. Monnow Bridge, Monmouth

HEW 146
SO 505 125

The Monnow Bridge at Monmouth was built in 1272 and is the only bridge in Britain which still carries a fortified tower.

The bridge, totalling 114 ft in length, has three semicircular masonry arches, each having three wide ribs. It has been widened by some 3 ft 6 in. on the upstream side and 5 ft on the downstream side to give a width between parapets of 24 ft. The widening has been carried out with great skill so that it blends well with the original structure.

Monnow Bridge

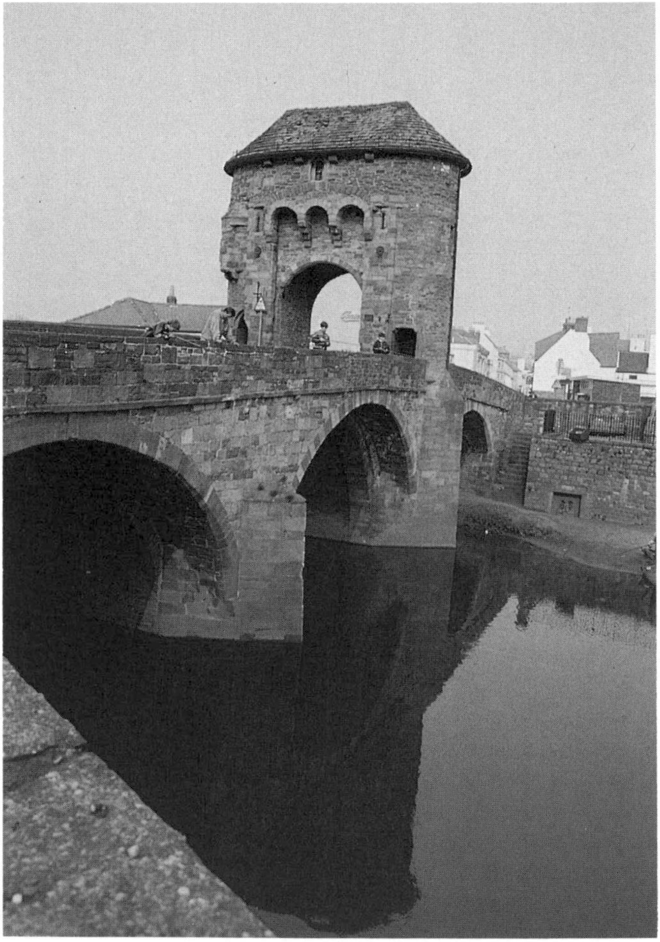

R. CRAGG

The fortification is arched over the roadway above the eastern pier of the bridge. Surmounting the archway there is a room 36 ft long by 10 ft wide covered by a pitched roof.

The bridge at Warkworth, Northumberland has a fortified tower at its southern end but not on the bridge itself.[1]

1. RENNISON R. W. *Civil engineering heritage: Northern England*. Thomas Telford, London, 1996, 18 (HEW 696).

2. Monmouth Forge Generating Station

HEW 1018
SO 503 137

A short way along the A466 north of Monmouth, an unclassified road to the left leads towards Osbaston. About 300 yd along this road another turning to the left leads to a small industrial complex. Here is the site of one of the earliest of the public electricity supplies in Britain and one of the earliest stations in Britain to generate at high voltage, at a time when most municipal generation was at 600 V for direct transmission to tramway networks.[1,2]

First commissioned in 1899, it had three alternators, each 7 kW, 3000 V 60 cycles single phase a.c., driven by water turbines under the brick-built station, which supplied electricity for lighting Monmouth.

The station was on the site of the Monmouth forge and tinplate works where there had been an ironworks from as early as 1628. Three hundred yards upstream the forge weir in the River Monnow was raised to provide an enlarged reservoir from which water was led to the station along a canal.

Each alternator had a stand-by steam engine in case of drought and a further steam engine (replaced by oil in 1923) drove a separate 21 kW generator. The system suffered initial teething troubles from lack of experience in the design of the several transformers scattered about the town to reduce the voltage to the consumers' voltage of 100/110 V. Originally established by the municipality, the station was sold to a private firm in 1930.

After nationalization of the industry in 1948, power continued to be fed to the national grid until 1953. The reservoir and weir[3] are still in use as a river regulator and can be reached by way of a public footpath across the field. Of the rest of the works only the station building remains, housing a small factory.

1. Tucker D. G. Half a century of hydroelectricity at Monmouth. *J. Monmouthshire Local History Council*, 1974, **37**, Spring, 27–37.

2. Harris P. G. *Wye valley industrial history*. Privately printed, 1976, 33–38.

3. Monmouth Power Station Weir (HEW 1269) SO 502 138.

3. Pant y Goitre Bridge

HEW 704
SO 348 089

According to Jervoise,[1] Pant y Goitre Bridge was built in about 1821 and locally it is considered that the engineer was probably John Upton, to whom is also attributed the nearby Llanellen Bridge at SO 306 111. It carries the B4598 road south of Llanvihangel Gobion over the River Usk.

The bridge has a centre span of 58 ft and side spans of 39 ft. The arches are semi-elliptical in form. The spandrel faces are pierced by cylindrical voids close to the springings and between the springings there is a larger void, 9 ft in diameter, over each pier. Similar 9 ft diameter voids through each abutment serve for flood relief.

Pant y Goitre Bridge

R. CRAGG

This is a most attractive bridge, graceful in shape and in beautiful surroundings.

1. JERVOISE E. *The ancient bridges of Wales and western England*. Architectural Press, London, 1936; republished by E.P. Publishing, East Ardsley, Wakefield, 1976, 106.

4. Crickhowell and Llangynidr Bridges

Just south of the A40 between Crickhowell and Bwlch two rather splendid sixteenth–seventeenth century masonry bridges cross the River Usk.

HEW 1223
SO 214 181

That at Crickhowell, which is a scheduled ancient monument, has 13 flat segmental arches ranging in span from 16 ft 10 in. to 39 ft 6 in. It was widened on the downstream side early in the nineteenth century but can still only carry single-line traffic. Its size and its position on the A4077, which turns south-west off the A40 through the village, suggest that the main coaching road once crossed it to take a route close to that adopted for the canal between Brecon and Abergavenny.

HEW 1247
SO 151 202

Five miles west of Crickhowell, the Llangynidr Bridge carries the B4560 road, about ½ mile south of its junction with the A40. The six segmental arches vary in span from 22 ft to 30 ft 6 in. The arch rings have two courses of

R. CRAGG

Crickhowell Bridge

R. CRAGG

Llangynidr Bridge

voussoirs, the outer course being shallower than the inner but overhanging it somewhat. The roadway is only 8 ft wide.

The bridge stands in a well-frequented beauty spot and, though shorter than Crickhowell Bridge, it is the more impressive of the two, with its massive piers and its arches standing higher above the river. It is a Grade II listed building.

As at Crickhowell, the bridge has triangular cutwaters to the piers, carried upwards to the tops of the parapets to provide pedestrian refuges.

5. Brynich Aqueduct

HEW 337
SO 079 273

Brynich Aqueduct is a four-span masonry structure carrying the Brecknock and Abergavenny Canal over the River Usk, just south of the A40 road, about 2 miles east of Brecon. It was built in 1799–1800, the Engineer being Thomas Dadford, junior.

It is approximately 210 ft in length. The canal width is 12 ft within an overall width of 31 ft and the water level of the canal is about 35 ft above river bed level. The spans are approximately 36 ft and the pier widths 7 ft 6 in.

6. Abercamlais Suspension Bridge

HEW 1265
SN 965 290

In the grounds of the Abercamlais estate adjoining the A40 road about 6 miles west of Brecon, a slender suspension bridge of 80 ft span crosses the River Usk. Built circa 1830, it is attributed to Crawshay Bailey, the famous iron master of Nantyglo in the eastern Ebbw Vale.

Two 1⅛ in. diameter wrought iron rods on each side, spaced 2 ft 8 in. apart vertically, form the suspension cables, their screwed ends passing through cast-iron end posts, 6 ft high and spaced 2 ft 2 in. at the bottom and 3 ft 4 in. at the top, to which they are tensioned by nuts. Every 10 ft, iron rods—hooked over the top cables and looped around the lower—carry transverse 4 in. by ⅜ in. flats as deck-bearers. Between these, short hangers attached to the lower cables carry 1½ in. by ½ in. bearers. The footway is of four 4 in. by ⅜ in. iron flats, spaced to give a width of 1 ft 6 in. and riveted to the wide deck-bearers but resting freely on the others. The bridge has been kept in excellent condition by the owners.

At Rhosferrig[1] near Builth Wells there was a similar but somewhat larger bridge with three suspension cables on each side, also attributed to Crawshay Bailey. Originally built over the Usk near Crickhowell about 1830, it was transferred to its later site over the Irfon in 1836. Unfortunately through age and flood damage it deteriorated beyond repair and was replaced in 1985 by the county council.

Just upstream from the Abercamlais bridge is a four-span arched masonry bridge which carries an estate road across the River Usk. It was built contemporaneously with the house as a packhorse bridge in about 1600, and widened a century later to carry wheeled traffic.

1. Rhosferrig Bridge (HEW 1184) formerly at SO 033 515.

7. Llandeilo Bridge

HEW 681
SN 627 220

Llandeilo Bridge has a clear span of 145 ft and an overall width of 33 ft. It was built in 1848 and carries the trunk road A483 on a falling grade from the town over the Afon Tywi and its neighbouring water meadows. The single masonry arch has a rise of 35 ft.

R. CRAGG

Llandeilo Bridge

The voussoirs are narrow but elongated and extend some way up the spandrel face. They are dressed to an ashlar finish with chamfered edges. The spandrel faces of the bridge, the arch soffit and the parapets, pilasters and wing walls are of squared, coursed masonry, with the faces chiselled or hammer dressed to an approximately flat surface. Pilasters occur at each end of the bridge and at intervals along the approach walls.

Perhaps in style it is too formal for such a rural setting but it has been described as probably the finest one-arch bridge in Wales. There are interesting references in *The Diary of Thomas Jenkins of Llandeilo* to its construction, on which the diarist worked.[1]

1. *Diary of Thomas Jenkins of Llandeilo 1826–1870*. Dragon Books, Bala, 1976.

8. Usk Dam

HEW 1240
SN 833 288

Not far from the source of the River Usk is an earth dam some 100 ft high and 1600 ft long on the crest, built in 1955 to create a water supply reservoir. It was designed by Binnie Deacon and Gourlay and built by Richard Costain Ltd. It is of historical interest because it was the first earth dam in Britain to be provided with horizontal drainage blankets in the embankment. This practice, suggested by

Professor A. W. Skempton of Imperial College, London, has since become standard when clay fill is used.

9. Upper Neuadd Dam

HEW 1241
SO 030 187

There are six reservoirs on the Taf Fawr and Taf Fechan rivers which flow from the Brecon Beacons to Merthyr Tydfil to form the River Taff. One of these is the Upper Neuadd, built in 1902 to the design of G. F. Deacon. The dam is of the gravity type, 77 ft 5 in. high and 1385 ft long on the crest, and is unusual in that the entire dam is of a thin masonry section buttressed with earth fill over most of its length.

10. Nant Hir Reservoir

HEW 1305
SN 989 068

Another tributary of the Taff, the Afon Cynon, has several interesting reservoirs on it above Aberdare. One of these, built in 1875 on a side stream at Nant Hir, has an earth dam some 63 ft high and 328 ft long on the crest, with a puddled clay core, a grassed downstream slope and stone pitched upstream slope.

Nant Hir Reservoir with Heads of the Valleys Road

R. CRAGG

It was the work of J. F. La Trobe Bateman, one of the most active water engineers of his day.[1] Other examples of his work are mentioned in Chapters 5 and 8. He was the son-in-law of William Fairbairn, whose work on tubular bridges in association with Robert Stephenson is referred to in Chapter 1.

Across the Nant Hir reservoir is a notable reinforced concrete open-spandrel arch bridge carrying the A465, the Heads of the Valleys Road.

1. BINNIE G.M. *Early Victorian water engineers*. Thomas Telford, London, 1981, 157–201.

11. Twrch Aqueduct

HEW 2074
SN 773 092

When the Swansea Canal was being built towards its final termination at Abercraf,[1] it was necessary to carry the waterway across the Afon Twrch, a tributary of the River Tawe at Ystalyfera.

The Engineer, Thomas Sheasby, designed a three-span masonry aqueduct with a channel 11 ft wide and 4 ft deep. To waterproof the channel he used hydraulic lime concrete, reputedly the first use of this material for this purpose. The segmental arches have spans of about 30 ft

Twrch Aqueduct, Swansea Canal

R. CRAGG

and the two piers have cutwaters on both sides. The cutwaters on the downstream side have a conventional triangular shape and originally the upstream cutwaters were of similar construction. However, because of the damage caused by boulders brought down by the Afon Twrch in times of flood, the upstream cutwaters were later rebuilt into the substantial curved structures seen today.

Immediately below the aqueduct the river falls over a stone weir about 13 ft high. This weir served not only to protect the foundations of the piers from scour but also diverted water into a feeder channel to the canal on the south-west side.

The aqueduct has a central spillway on the upstream side and stop plank grooves at both ends. A sluice on the upstream side of the channel enables the aqueduct to be drained of water for maintenance.

On the north-east side of the aqueduct is a culvert which carried water from the tail race of a nearby fulling mill.

The Twrch Aqueduct was completed in 1798. Although this section of the Swansea Canal has now largely vanished, the aqueduct has been preserved.

1. RUSSELL R. *Lost canals of England and Wales.* David and Charles, Newton Abbot, 1971, 128–130.

12. South Wales Railway

HEW 1199
ST 536 937 to SN 412 196

The Great Western Railway (GWR) broad gauge (7 ft ¼ in.) lines were extended into South Wales in 1851 via Gloucester and Chepstow.

Apart from the bridge over the Wye at Chepstow, notable structures on the South Wales Railway were the crossing of the Usk at Newport, subsequently rebuilt and widened; Landore Viaduct,[1,2] on which many timber spans were replaced by iron in 1888 and by steel in 1979; Llwchwr Viaduct,[3] another timber viaduct whose deck is now in steel; and a drawbridge opening span at Carmarthen, built by Brunel in 1854 and rebuilt in 1910 as a Scherzer rolling lift bascule bridge.[4,5] Models of the main span of the original Landore Viaduct are in the Swansea

Maritime and Industrial Museum and the National Railway Museum in York.

The curves at Neath are not conducive to high-speed running and the few stiff gradients thereabouts made heavy demands on motive power. This situation was eased in 1913 by the construction of a bypass line which crossed the River Neath by an interesting swing bridge,[6] originally built by the Rhondda and Swansea Bay Railway. Five fixed spans of plate girder construction, varying in length from 40 ft to 52 ft 6 in., have been renewed. The movable span is a Pratt truss with a curved upper boom, swinging about its centre and resting on a cast-iron roller race, the span being 167 ft 6 in. Originally operated hydraulically, the bridge is now fixed. It is the only opening bridge of this type in Britain built both on the skew and on a curve. Finch of Chepstow supplied the steelwork; Sir William Armstrong of Newcastle the hydraulic machinery.

At Llansamlet, between Neath and Swansea, four 70 ft span stone arches,[7] which may be compared with the Chorley flying arches,[8] spring from the sides of a cutting. Brunel seems to have built them to permit steeper slopes and so save on excavation.

Bridgend station was overhauled in 1980 and is now a pleasant mixture of the original Brunel buildings and modern railway architecture, which received awards from HRH the Prince of Wales and from the Development Corporation of Wales.

The South Wales Railway was originally intended to lead to a cross-channel port at Fishguard but economic conditions in Ireland at the time defeated this object, and Fishguard was not developed seriously until 1904.

The use of the broad gauge after it had been officially declared non-standard in the United Kingdom was unfortunate, since the huge output from the South Wales collieries demanded mineral wagons which would be capable of working, not only over any part of the English rail network irrespective of track gauge, but also up the Welsh valleys themselves in an economic manner. Even Brunel had thought in 1840 that something smaller than the broad gauge was desirable there when he built the Taff Vale Railway. By 1866, so great was the clamour in

South Wales to convert the main land exit route eastward into England to standard gauge that this was achieved by 1872.

The long detour via Gloucester was still a disadvantage for London, Bristol and West of England traffic. Brunel's original proposals had indeed envisaged a crossing of the Severn at Hock Cliff (SO 730 090) but this was vetoed by the Admiralty. Later a more direct link near Chepstow was opened, but it was not until 1886 that the GWR solved the problem by the construction of the Severn Tunnel and, for London traffic, by a direct link from it to Swindon in 1903.

1. Landore Viaduct (HEW 327) SS 663 959.

2. FLETCHER L. E. Description of the Landore viaduct on the line of the South Wales Railway. *Min. Proc. Instn Civ. Engrs*, 1854–55, **14**, 492–506.

3. Llwchwr Viaduct (HEW 1131) SS 561 980.

4. Carmarthen Railway Bridge (HEW 1216) SN 405 192.

5. WHITLEY H. S. B. The reconstruction of Carmarthen bridge. *Min. Proc. Instn Civ. Engrs*, 1916–17, **204**, 365–368.

6. Neath River Swing Bridge (HEW 1232) SS 730 964.

7. Llansamlet Arches (HEW 802) SS 702 975.

8. RENNISON R. W. *Civil engineering heritage: Northern England*. Thomas Telford, London, 1996, 219–220 (HEW 751).

13. Penydarren Tramroad

HEW 705
SO 056 070 to ST 085 950

The Penydarren Tramroad, engineered by George Overton and opened in 1802, has a unique place in railway history. On it, on 21 February 1804, what is generally regarded as the first journey by a steam locomotive was made, using a machine built by Richard Trevithick. The L-section cast-iron rails were 3 ft long, weighed 56 lb each, and were laid to a gauge of 4 ft 4 in. over the outside of the flanges. The rails proved too fragile for the weight of the locomotive, so that although the experiment showed the possibilities of this new form of traction, it also demonstrated that these could not be fully realized until a much better track became available.

The Welsh Industrial and Maritime Museum at Cardiff has a working replica of the locomotive. The rails upon which it runs are replicas of the Penydarren Tramroad rails but laid on concrete blocks instead of stone. A

O. M. GIBBS/WELSH INCUSTRIAL AND MARITIME MUSEUM

Penydarren Tramroad: replica of Trevithick's locomotive

modification to the seating of the rails on the blocks has, by reducing the effective span, overcome the breakage problem.

The Glamorganshire Canal was completed between Merthyr Tydfil and the sea at Cardiff in 1798. In consequence of disagreements between Richard Crawshay, whose Cyfarthfa ironworks were best served by the canal, and the other ironmasters of Merthyr, a tramroad was constructed from Samuel Homfray's Penydarren works to near a point on the canal at Abercynon (the nearby Navigation Hotel was once the offices of the canal company). This route effectively bypassed part of the canal route, particularly a set of ten locks north of the junction at Abercynon.

From its southern terminus at Abercynon (ST 085 950) the route of the tramroad runs along the eastern bank of the River Taff. Near Quaker's Yard the river was crossed by timber bridges[1] at ST 094 963 and ST 090 966. In about 1815 these were replaced by segmental, nearly semicircular, masonry arches of 60 ft span and 9 ft wide, which have parallel rings about 18 in. deep composed of 2 to 3 in. flat stone voussoirs similar to Pont y Gwaith, and indeed, Pontypridd. The old alignment can still be traced.

Further on, the tramroad was eventually crossed by the Taff Vale Railway on a viaduct.[2] The section between Quaker's Yard and Pont y Gwaith follows the deep valley of the Taff through pleasantly wooded country.

North of Merthyr Vale the route crosses to the east side of the A470 trunk road and continues to Merthyr where its course is perpetuated in the names of two streets, Tramroadside South and Tramroadside North.

At SO 056 070, near the point of termination, there is a memorial to Richard Trevithick. The southern portal of a tunnel at SO 058 045 was exposed during land reclamation work at the site of the former Plymouth Ironworks in the 1980s and has been preserved.

1. Penydarren Tramroad Bridges (HEW 799) ST 090 966 and ST 094 963.

2. Quaker's Yard Viaduct (HEW 801) ST 089 965.

14. Taff Vale Railway

HEW 1219
SO 052 057 to ST 190 750

At Pontypridd the River Taff is joined by the River Rhondda before flowing on to Cardiff. Between the two rivers lies the valley of the Cynon, a tributary of the Taff.

At the turn of the eighteenth and nineteenth centuries

the discovery of iron ore along the head of the valleys and coal in the valleys led to the establishment of ironworks, as at Cyfarthfa and Dowlais near Merthyr Tydfil, and to a rapid expansion in coal mining. These industries required means of transport for the raw materials and for the finished products. The Glamorganshire Canal from Merthyr to Cardiff was built between 1790 and 1798 to give communication with the sea, but in general the South Wales countryside did not really lend itself to the development of canals and so there was a great proliferation of horse tramroads such as Penydarren.

By 1830, canals and tramroads were proving inadequate, so in 1835 the Merthyr ironmasters engaged I. K. Brunel to construct the first major commercial railway in South Wales. An Act of June 1836 authorized a single 4 ft 8½ in. gauge track, 24¼ miles long with six passing places, from Merthyr, at the head of the Taff Vale, to Cardiff, where dockland development was beginning.

Major works included a rope-worked 1 in 19 incline north of Navigation (Abercynon); a masonry viaduct[1] at Pontypridd with a skew span of 110 ft across the Rhondda; the six-span masonry viaduct across the Taff,[2] and the Penydarren Tramroad at Quaker's Yard. This is unusual in having octagonal piers with deeply chamfered arches. The line was opened in 1840–41.

In the course of time, coal became the dominant traffic. The Taff Vale Railway expanded into the Cynon valley from 1845 and by 1856 it had reached Treherbert near the head of the Rhonnda Fawr. Before the end of the century the company had penetrated to Llantrisant, Cowbridge and Aberthaw and beyond Cardiff to Penarth, where it eventually owned the docks; and to Barry, which gave the railway satisfactory outlets to the sea.

The Taff Vale, being first in the field, was able to choose the easiest routes in difficult country. Later railways competing for the lucrative coal traffic had to contend with harsh topography, sometimes having to move from valley to valley, and at the expense of long viaducts and tunnels.[3] It is little wonder, then, that the surviving railway lines in the area largely use the pioneering routes of the Taff Vale Railway.

1. Pontypridd Railway Viaduct (HEW 1220) ST 071 900.

2. Quaker's Yard Viaduct (HEW 801) ST 089 965.

3. ATKINS W. S. Transportation—road and rail. *Civil engineering problems of the South Wales valleys*. Institution of Civil Engineers, London, 1970, 41–54.

15. Pontypridd Bridge

HEW 27
ST 074 904

No study of bridges—especially masonry arch bridges—would be complete without a reference to William Edwards' famous arch across the River Taff at Pontypridd.[1,2] His first attempt, in 1750, a multi-span bridge, was washed away in a flood after only two years. Edwards then decided to span from bank to bank, but again his arch was washed away in a flood even before the centre was struck. His next attempt was also doomed to failure. The span of 140 ft involved a rise of 35 ft, which meant a great weight of filling over the haunches compared with the crown where there was only the arch ring and the parapets. During the construction of the spandrel walls the excessive weight near the abutments forced the crown upwards and the bridge again collapsed. Fortunately this was not a sudden failure and Edwards had time to observe the mode of collapse.

Jervoise[3] has described the third and successful single

Pontypridd Bridge

INSTITUTION OF CIVIL ENGINEERS

span as follows: 'Edwards then rebuilt the bridge to the same design except that he placed at each end three cylindrical holes graduated in size, the largest being 9 ft in diameter, to relieve the arch from the pressure of its haunches.' The spandrel infilling was of charcoal, for further lightness. This scheme proved successful and the bridge, which was completed in 1755, still stands.[4]

The bridge soffit is an almost perfect arc of a circle of 89 ft radius and the arch ring has a depth of construction of only 2 ft 6 in. The relatively large rise at the crown resulted in steep slopes at either end of the bridge and this caused serious problems for heavy carts, both during the ascent and descent. A modern bridge has been built alongside and Edwards' masterpiece is preserved for use by pedestrians only. It is 11 ft wide between parapets.

1. SMITH T. M. Account of the Pont-y-tu-prydd over the River Tâfe near Newbridge in the county of Glamorgan. *Min. Proc. Instn Civ. Engrs*, 1846, **5**, 474–477.

2. BRUCE G. B. The Royal Border bridge. *Min. Proc. Instn Civ. Engrs*, 1850–51, **10**, 241–242.

3. JERVOISE E. *The ancient bridges of Wales and western England*. Architectural Press, London, 1936; republished by E.P. Publishing, East Ardsley, Wakefield, 1976, 93–94.

4. BELL W. On the stresses in rigid arches, continuous beams and curved structures. *Min. Proc. Instn Civ. Engrs*, 1871–72, **33**,58–63.

16. Berw Road Bridge, Pontypridd

HEW 620
ST 077 911

The Berw Road Bridge crosses the River Taff about half a mile upstream from William Edwards' masonry arch bridge. It is one of several early reinforced concrete bridges in South Wales built on the system developed by the well-known French engineer François Hennebique, who lived in South Wales for part of his life and was for a time the French consul for the area.

The bridge has a central clear span of 116 ft and side spans of 25 ft. The width between parapets is 26 ft. The main span has three parabolic arched ribs at 12 ft centres, braced at intervals. The longitudinal beams supporting the deck are supported by columns off the ribs over the outer thirds of the span, and the arch itself serves as direct support to the deck over the middle third. The side spans

O. M. GIBBS

have their outer main beams arched to match the centre span.

Berw Road Bridge

The bridge was completed in 1909 to the design of L. G. Mouchel and Partners. The deck was reconstructed in the 1970s.

17. Pont y Gwaith

HEW 800
ST 080 975

Pont y Gwaith is a masonry bridge of 55 ft span, 15 ft 9 in. rise, over the River Taff about a mile north of Quaker's Yard. It has several features in common with Pontypridd, including the use of thin stones to form the arch ring, the severe road gradient, and the narrowing in plan from the abutments to mid-span, but it has no opening in the spandrels. Both span and width of the bridge are much smaller than those of Pontypridd. There is a noticeable tendency of the arch to be pointed at the crown like the first Ouse Bridge at York.

Following damage by mining subsidence, the structure was rebuilt in 1993 and now forms part of the Taff Leisure Trail.

R. CRAGG

Pont y Gwaith

18. Cefn Coed y Cymmer Viaduct

HEW 171
SO 030 076

During the second half of the nineteenth century the network of railways expanded rapidly in South Wales to handle the vast tonnages of coal being produced. The difficult terrain forced the construction of numerous viaducts, some with masonry arches, some with metal spans. Often of great height and curved in plan, they produced some excellent examples of bridge engineering. Many have since been demolished and others are disused, such as Cefn Coed, a masonry viaduct which carried a part of the Brecon and Merthyr and London and North Western Joint Railway over the Taf Fawr. Built on a curve, it is 770 ft long and 115 ft high with 15 semicircular arches of 39 ft 9 in. It was designed by Alexander Sutherland in consultation with Henry Conybeare and built by Savin and Ward in 1866.

The line opened on 1 August 1867 and closed in 1962.

19. Pont y Cafnau

HEW 656
SO 038 071

This unique cast-iron 'bridge of troughs' still spans the River Taff where it was built in 1793 to carry a tramroad and water supply into the Cyfarthfa ironworks in Merthyr Tydfil.[1] The designer was the Chief Works Engineer, Watkin George. In 1795 a second bridge, which no longer exists, was cast from the same patterns to carry an extension of the tramroad from the works to the Glamorganshire Canal.

The bridge, now used by pedestrians, spans 47 ft. Two substantial A-frames, one on each side of the deck, have their feet embedded in the river walls, with the apex at mid-span. The frames are held together by mortise and tenon and dovetail joints (George was a former carpenter) and incorporate sockets which carry transverse members at mid-span and at the quarter points. These in turn support the deck structure, which is a closed rectangular box about 2 ft deep and 6 ft 2 in. wide.

Pont y Cafnau undoubtedly had its influence on other, better known, aqueducts. In 1794 the Shropshire iron master, William Reynolds, sketched the bridge and in the following year Telford reported that the design and

Cefn Coed y Cymmer Viaduct

R. CRAGG

method of construction at Longdon on Tern Aqueduct, itself a prototype for Pontcysyllte, had been referred to Reynolds and himself.

In recent years the structure has been refurbished by the local authority and this has exposed an interesting arrangement whereby the rail chairs are cast integrally with the deck.

1. HAGUE D. and HUGHES S. Pont y Cafnau, the first iron railway bridge and aqueduct? *Ass. Industrial Archaeology Bulletin*, 1982, **9**, No. 4, 3–4.

Pont y Cafnau

R. CRAGG

Cwmbach
Railway Bridge

20. Railway Bridge at Cwmbach, Aberdare

HEW 1055
SO 024 011

This steel truss bridge, which carries the only remaining railway line in the Aberdare valley over the Afon Cynon, played an interesting role in the development of modern design practices and construction techniques.

In the 1960s, the A40 trunk road east of Oxford was being widened from single to dual carriageway, and at Wheatley a bridge was erected in 1961 to carry the single-track Oxford to Thame branch over the new road. It was required to match in general appearance the adjacent riveted truss bridge over the existing road.

P. S. A. Berridge was responsible for the design and

chose to develop a structure which was largely prefabricated by welding, but with the necessary degree of site assembly carried out by Torshear bolts, which acted by applying a known pressure to the mating surfaces of the joints.

The experience gained was used in the construction of the much larger trusses, also designed by Berridge, which, in 1962, replaced I. K. Brunel's tubular suspension bridge at Chepstow. Both bridges were fabricated at the Fairfield works at Chepstow, successors to Messrs Finch, who did so much work for Brunel.

In 1973, on the closure of the Thame branch, the bridge at Wheatley was dismantled and reassembled near Aberdare. This time the more recently developed Huck fasteners were used for joint assembly.

21. Robertstown Cast Iron Tram Bridge, Aberdare

HEW 371
SN 997 037

This interesting little bridge is certainly one of the oldest surviving 'railway' bridges in the world. In 1811 the Aberdare Canal Company completed a tramway between Hirwaun and the canal head at Cwmbach. The bridge carried the tramway across the River Cynon between Trecynon and Robertstown.

Robertstown Tramway Bridge

R. CRAGG

Four arched and trussed cast-iron beams spring from continuous cast-iron brackets built into the abutments. The width of each truss is only 3 in. and the depth varies from 1 ft at the centre to over 5 ft at the ends. Seventeen cast-iron plates 9 ft 11 in. wide make up the total length of deck of 36 ft 8 in. The stone abutments were built with obvious skill.

The bridge is now used as a footway.

22. Hengoed Viaduct

HEW 804
ST 155 949

The masonry Hengoed, or Maes-y-Cwmmer Viaduct, like the iron viaduct at Crumlin, was one of the major structures on the Taff Vale line of the Newport, Abergavenny and Hereford Railway and was built in 1857. It has 16 spans of 40 ft semicircular arches and a maximum height of 130 ft. It is built in rough stone, on a curve, and crosses the A469 (spans three and four), the River Rhymney (span seven) and a minor road (span eleven).

It is now planned for the viaduct to be repaired and included in the route of a cycle-way linking Newport to the Rhondda.

23. Carew Tide Mill

HEW 803
SM 041 038

Carew Tide Mill, a large three-storey stone building, stands at the southern end of a dam across the estuary of the Carew River near the southern edge of Milford Haven. Records of the mill can be traced back to Elizabethan times.

The dam, which has stone facings and a central clay core, impounds an area of water of some 27 acres and this water is discharged through passages in the dam beneath the mill to operate undershot water-wheels, of wood and iron construction. Only one wheel is now in operation and it is considered to generate some 20 brake horse power.

After a long period of disuse the mill was saved from imminent collapse in the early 1970s, largely through the efforts of Mr John Russell FSVA, and with financial assistance from various organizations. Following restoration the mill now houses an interesting milling museum.

J. P. DAVIES

Carew Tide Mill

In the 1930s the sites of some 25 tide mills were recorded in southern Britain. Today, few are left and even fewer, among which may be mentioned that at Woodbridge in Suffolk,[1] are still in working order.

1. LABRUM E. A. *Civil engineering heritage: Eastern and Central England.* Thomas Telford, London, 1994, 147 (HEW 113).

24. Three Railways near Llanelli

The three railways described here were all based on earlier transport systems, the first being on the route of a canal and the other two following the lines of old horse tramways.

In 1766, Thomas Kymer obtained powers for a 3 mile canal from Kidwelly Harbour (the first Welsh canal to obtain an Act of Parliament)[1] and this canal eventually extended northward up the valley of the Gwendraeth Fawr and eastward to Burry Port. At Burry Port a tramroad led further east to a junction with the Carmarthenshire Railway near Llanelli. In 1865 the canal company obtained powers to abandon the canal and construct a railway in its bed and in 1866 it amalgamated with the Burry Port and Gwendraeth Valley Railway Company which had been established in the same year.[2] Sir William Shelford and John Robinson were the Engineers for the conversion. The railway line followed the route of the canal and its tramway from Kidwelly Harbour, passing under the Great Western Railway (GWR) South Wales Railway by a low bridge before running northward to a terminal station at Cwmmawr. From a triangular junction at SN 428 059, the route to Burry Port and Llanelli ran eastward, crossing the GWR line again at Burry Port. Low bridges were a characteristic of the line, betraying its canal ancestry, for many of the original canal bridges were not modified for the railway and specially cut-down locomotives have always been necessary to work the traffic, even in recent times when modified class 08 diesel shunters were used. Passengers were carried from 1909 when a Light Railway Order was obtained, and the line was absorbed by the Great Western Railway in July 1922.

HEW 1217
SN 451 008 to
SN 529 125

The railway made use of two old canal aqueducts, the one over the Gwendraeth Fawr on the Llanelli line having six arches of about 9 ft span.[3] The second aqueduct is at SN 447 073, near Pont Newydd.

Passenger services on the line were withdrawn in 1953 and general freight traffic ceased in 1965. Today only the line from the GWR connection at Kidwelly to Cwmmawr survives, having until recently carried coal from the Cwmmawr opencast disposal point, but recent reports state that even this traffic ceased in March 1996 and at the time of writing the line is closed completely.[4,5]

HEW 1388
SS 500 995 to
SN 563 131

The second line, the Llanelli and Mynydd Mawr Railway (L&MMR), had its origins in the Carmarthenshire Railway or Tramroad, a horse-drawn line running from the harbour at Llanelli northward to limestone quarries near Castell-y-Graig (SN 603 158). It was only the second public railway to be authorized by an Act of Parliament, of 3 June 1802, the first having been the Surrey Iron Railway of 1801.[6] It had flanged rails on stone blocks. Never commercially successful, it was closed in 1844.

In 1877 the L&MMR was incorporated to lay a standard gauge railway on the old trackbed. It opened for freight traffic on 1 January 1883 and by 1887 extended 13 miles to Cross Hands. There was no passenger service but large numbers of colliery workers were carried in workmen's carriages. Originally the line crossed over the GWR at Llanelli to reach the docks with a connection to the main line railway at the crossing point, but the docks line was later taken up. In 1923 the L&MMR was absorbed by the GWR. After 1966 the line was not used north of Cynheidre Colliery (SN 495 075) and with the recent closure of this colliery the whole line is now disused.[7]

HEW 1389
SS 532 989 to SN 634 226

The final railway, the Llanelli Railway, is still, in part, in use as a section of the Central Wales line from Craven Arms in Shropshire to Llanelli and Swansea. Its origins lie in the Llanelli Railway and Dock Company which was incorporated in 1828 to make a wet dock at Llanelli and a horse tramroad. By 1839 it had built a railway from Llanelli to Pontardulais and this was extended in stages to reach Llandeilo in 1857. Here the Company leased the Vale of Towy Railway and so reached Llandovery in April 1858. Branches were opened into the Amman valley from Pantyffynnon in the 1840s and to Carmarthen in 1865, and in 1867 a further line from Pontardulais to Swansea opened, crossing the Gower and running along the shore eastward to Swansea (Victoria) station.

In 1868 the London and North Western Railway (L&NWR) Central Wales line reached Llandovery from the north and the L&NWR was granted running powers over the Llanelli Railway which enabled it to run into Swansea from January 1873. The Llandeilo to Carmarthen and Pontardulais to Swansea lines were absorbed by the L&NWR in 1891 and the remainder of the Llanelli Railway passed into the hands of the GWR in 1889.[8]

Only the original main line from Llandovery to Llanelli remains open, carrying a service of trains from Shrewsbury which reach Swansea along the GWR main line by a reversal of direction at Llanelli. A mineral branch from Pantyffynnon also remains in operation.

1. RUSSELL R. *Lost canals of England and Wales*. David and Charles, Newton Abbot, 1971, 131–133.

2. BARRIE D. S. M. *A regional history of the railways of Great Britain, Volume 12: South Wales*. David and Charles, Newton Abbot, 1980, 226–229.

3. Gwendraeth Fawr Bridge (HEW 1234) SN 428 054.

4. *The Railway Magazine*, 1996, **142**, 1142, June, 13.

5. *The Railway Magazine*, 1996, **142**, 1143, July, 20.

6. Otter R. A. *Civil engineering heritage: Southern England*. Thomas Telford, London, 1994, 180–181 (HEW 1387).

7. Barrie D. S. M. Op. cit. 224–226.

8. Barrie D. S. M. Op. cit. 210–212.

25. Whiteford Point Lighthouse

HEW 1256
SS 444 973

This has been described as the only operational wave-swept cast-iron lighthouse in the British Isles. A scheduled ancient monument, it stands a mile offshore on the north-west tip of the Gower Peninsula and looks across the Burry Inlet to Burry Port and Llanelli, whose harbours have now largely fallen into disuse. It was built in 1865 under the Llanelli Harbour Act of 1864 to replace an

N. J. FRANCIS, LLANELLI

Whiteford Point Lighthouse

earlier light, carried on timber piles, which had lasted only a few years. The light was extinguished in 1921, but in 1982, with the aid of a grant from the Llanelli Harbour Trust, the Burry Port Yacht Club installed a solar-powered light in the lantern.

The lighthouse stands just above the low water mark and is 61 ft high. Eight courses of cast-iron plates, bolted together through internal and external flanges, form the circular tower which has the curved taper typical of many lighthouses. The lantern is surrounded by an ornate wrought iron balcony. At some time after erection the tower was strengthened by wrought iron bands.

The area to the south of the lighthouse, which is some 2 miles from the nearest road, is in the care of the National Trust. The lighthouse is accessible on foot at low tide but because of unexploded shells and quicksands in the vicinity and the sensitive nature of the reserve, intending visitors should first get in touch with the Warden at the Oxwich Reserve Centre, Oxwich, Gower.

26. Swansea and Mumbles Railway

HEW 706
SS 645 921 to
SS 629 874

The Oystermouth Railway, as it was originally known, was authorized by Parliament in June 1804 and was the world's first fare-paying passenger 'railway'.[1,2] The first load of passengers was carried on 25 March 1807, using horse traction. The Act mentions haulage by 'men, horses or otherwise'—the 'otherwise' perhaps having Trevithick's Penydarren locomotive in mind.

Steam traction was introduced in 1877. In 1879 the line was renamed the Swansea and Mumbles Railway and in 1898 was extended to Mumbles Pier. It was electrified in 1928, being operated by double-decker tramcar type vehicles, the largest in use in Britain, and continued in service until 5 January 1960.

The track ran along the seaward side of the A4067 which links the Gower Peninsula to Swansea. Although the tracks have now been removed, the course of the 6 mile route can still be followed and the former Blackpill station, built in 1927, which also housed an electricity sub-station, remains in good condition.

In March 1981 a stained glass window was unveiled

in Oystermouth parish church to mark the 175th anniversary of the opening of the railway. A replica of the original horse carriage and the cab of the last Mumbles Railway tram can be found in the Swansea Maritime and Industrial Museum. The route of the railway can today be followed on the cycle-way running along the bay.

1. Mumbles Railway Society. *The Mumbles Railway, the world's first passenger railway*. R. Cottle, Swansea, 1981.

2. Lee C. E. *The first passenger railway—The Oystermouth or Swansea and Mumbles line*. Railway Publishing Co., London, 1942.

27. New Inn Bridge, Merthyr Mawr

HEW 1230
SS 891 784

New Inn Bridge takes its name from a hostelry that once stood near this bridge, which carried the main South Wales coaching route over the Ogmore river. This route is now a minor road just south of the A48 and 1 mile south-west of Bridgend. The bridge is of masonry, having four pointed arches. Cutwaters protect each pier and have stepped upper parts. Each span is about 15 ft and the width between the parapets is 10 ft.

The parapets contain openings through which sheep were once forced to leap into the deep pool beneath the

New Inn Bridge

R. CRAGG

bridge. For this reason the bridge, which is several hundred years old and a scheduled ancient monument, is also known as the Sheep Dip Bridge.

28. Glanrhyd Bridge

HEW 1084
SS 899 828

The Duffryn Llynfi tramroad linked ironworks and collieries in the vicinity of Maesteg to a small harbour at Porthcawl.

A branch from Tondu to Bridgend was carried across the Ogmore river by means of a three-span masonry bridge known as Glanrhyd or Tyn y Garn Bridge. The arches span 23, 31 and 26 ft and the width between parapets is 13 ft.

The bridge, on which a stone plaque records that it was built in 1829, is well proportioned and of good workmanship. It now carries a minor road which turns off the Bridgend to Maesteg road, the A4063, to the north of Pen y Fai hospital. The bridge lies below a span of the Ogmore valley viaduct of the M4 motorway and served to support a section of the formwork during the construction of the viaduct. The two structures in such close proximity vividly illustrate the change in scale of transport provision over the last 150 years.

Glanrhyd Bridge

O. M. GIBBS

The tramroad was built to a gauge of 4 ft 7 in. and was intended for horse-drawn traffic. In due course the lines were converted to standard gauge and from 1861 were worked by steam locomotives, although Glanrhyd Bridge itself was never used by these.

Commemorative plaques now mark the termini of the tramroad at Maesteg, Bridgend and Porthcawl.

29. Barry Docks

HEW 1233
ST 124 668

In 1889 the Barry Railway Company built a dock between Barry Island and the mainland to capture a share of the export trade in South Wales coal, which had developed enormously over the previous 50 years. The Engineer was John Wolfe Barry, one of whose assistants was I. K. Brunel's son Henry. T. A. Walker was the Contractor. One of the principal advocates of the scheme was David Davies, already a well-known railway contractor, the owner of the Ocean Collieries and the leader of the Rhondda mine owners. His statue stands in front of the former dock office at Barry.

The works comprised a basin 500 ft by 600 ft and a dock 3400 ft by 1100 ft divided by a mole into two arms at its western end. The wrought iron gates in the 80 ft entrance to the basin and between the basin and the dock were the first to be operated directly by hydraulic rams instead of through opening and closing chains. At the north-east corner of the dock there was a dry dock 740 ft by 100 ft with a 60 ft entrance closed by a caisson.

The dock was liberally supplied with hydraulic hoists for tipping coal wagons into ships. The excavation was generally left with sloping sides. Walls were provided only round the basin, along the south side of the dock and at the positions of the hoists; they are 46 ft 6 in. high from dock bottom to coping, and were built in the dry, using massive limestone blocks.[1]

Breakwaters were constructed off the entrance, mainly with rubble from the excavations, faced with 4 ton stone blocks. On the end of the west breakwater is a cast-iron lighthouse 30 ft high, 7 ft 9 in. diameter at the base and 6 ft 6 in. diameter at the top, which may be compared with Whiteford Point Lighthouse, described earlier in this chapter.

Before the works started, coffer-dams had to be formed by tipping across the channel to enclose the dock site and to enable the 5 million cubic yards of excavation to proceed. A Cornish engine of 267 000 gallons per hour capacity was brought from Walker's Severn Tunnel works to assist in the dewatering.

A pier was built on the Bristol Channel side of the docks, with a passenger line from Barry Island station to the Pierhead station through a tunnel, now closed, to the east of the holiday camp.

A second dry dock, similar to the first, was opened in 1893.[2] The 34 acre No. 2 Dock was opened off No. 1 Dock in 1898; while in 1908 the Lady Windsor Lock, 647 ft long and 65 ft wide, and named after the wife of the company chairman, was opened directly into the sea to the west of the tidal basin.

By the end of the nineteenth century the coal trade had reached its peak and after a few years entered upon a continuous decline. Today, no coal is handled at the docks, but the present owners, Associated British Ports, have adapted them to handle a variety of trades, including the export of coke.

Some general notes on the South Wales docks will be found in the introduction to this chapter.

1. Robinson J. The Barry Dock works, including the hydraulic machinery and the mode of tipping coal. *Min. Proc. Instn Civ. Engrs*, 1889-90, **101**, 129–151.

2. Robinson J. The Barry Graving Docks. *Min. Proc. Instn Civ. Engrs*, 1893–94, **116**, 267–274.

30. Melingriffith Water Pump

HEW 392
ST 143 800

The Glamorganshire Canal, which was opened in 1798, drew water from a feeder which also supplied the Melingriffith Works in the Whitchurch area of Cardiff. The Canal Company was obliged to take its waters from the tail race below the Melingriffith Works, and because of this a pumping engine had to be installed to lift the water 12 ft into the canal feeder. Designed and constructed by John Rennie and William Jessop, it operated for 135 years until the closure of the canal in 1942. The pump was well

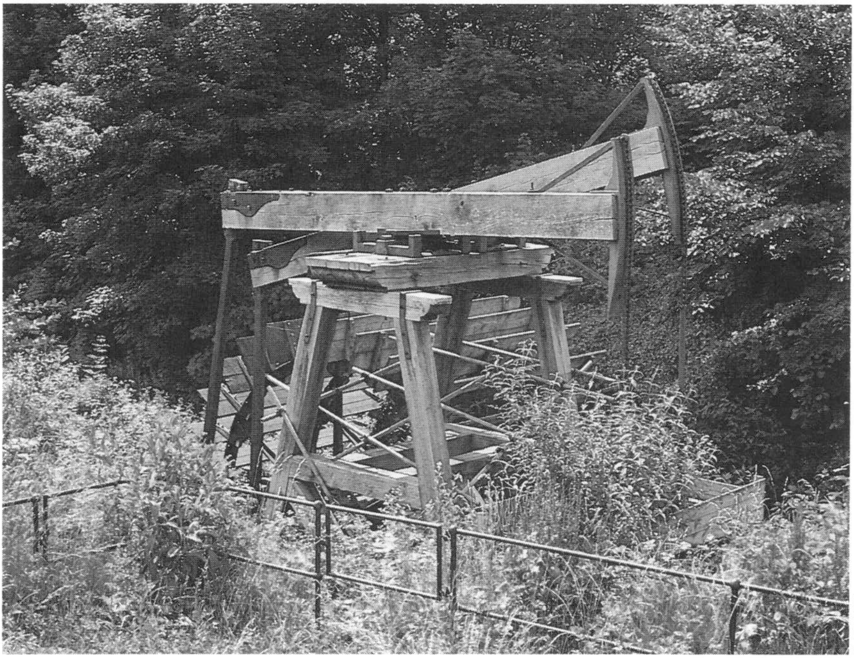

constructed of American oak and cast iron and survived until restoration work started in 1974.

Melingriffith Water Pump

The undershot water-wheel is 18 ft 6 in. in diameter by 12 ft 6 in. wide and consists of three cast-iron wheels mounted on an axle, originally of oak, now of steel. There are six cruciform cast-iron spokes to each and the rims carry 30 paddles, 22 in. deep. The wheel drove pistons in two cylinders via timber rocking beams 22 ft long. The cylinders had a bore of 2 ft 8 in. and a 5 ft stroke.

There is a model of the pump in the Department of Industry at the National Museum of Wales, Cathays Park, Cardiff.

31. Crumlin Viaduct

HEW 72
ST 213 986

Of all the many viaducts in South Wales, perhaps the most interesting and certainly the most dramatic was that at Crumlin.[1,2] It is still worth describing although it was demolished in 1966. There is a small model in the

National Museum of Wales and traces of the piers can be found on site.

Crumlin Viaduct had ten spans, divided into groups of seven and three by a rock knoll. The total length approached 1700 ft. Crossing the valley of the Ebbw river at a height of 200 ft, the viaduct was built in 1857 by the Newport, Abergavenny and Hereford Railway for their line to the Taff Vale. The piers each consisted of twelve cast-iron columns 12 in. diameter in three rows of four, with an additional raker at each end, covering an area of about 42 ft by 21 ft. Each column was in eight pieces with wrought iron cross bracing to adjacent columns. There were four lines of Warren truss girders, each of 150 ft span and 15 ft deep, cross braced by iron plates in 1868 and by steel from 1930. This was the first large-scale multi-span use of this design in wrought iron.

The viaduct was designed and built by T. W. Kennard, the partner of James Warren, who devised the truss configuration which bears his name, and which was first used for a major bridge at Newark Dyke on the Great Northern Railway main line.[3,4] The Crumlin columns were cast at Kennard's works and the wrought iron, supplied by the Blaenavon Works, was fabricated on site and hoisted into position.[5]

Crumlin Viaduct

METIUS CHAPPEL: *BRITISH ENGINEERS* (I.C.E.)

1. MAYNARD H. N. *Handbook to the Crumlin Viaduct*. Wilson, Crumlin, 1862.

2. HUSBAND J. The aesthetic treatment of bridge structures. *Min. Proc. Instn Civ. Engrs*, 1900–01, **145**, 217.

3. Newark Dyke Bridge (HEW 1023) SK 801 558.

4. CUBITT J. A description of the Newark Dyke Bridge on the Great Northern Railway. *Min. Proc. Instn Civ. Engrs*, 1852–53, **12**, 601–612.

5. MAYNARD H. N. Discussion: The different modes of erecting iron bridges, by T. Syrig. *Min. Proc. Instn Civ. Engrs*, 1880–81, **63**, 189–191.

32. Flight of Locks near Rogerstone, Monmouthshire Canal

HEW 798
ST 286 884 to
ST 279 886

The most spectacular set of locks in South Wales is at Rogerstone near Newport on the Crumlin branch of the Monmouthshire Canal, north of the M4 motorway between junctions 26 and 27. There are 14 locks in all, only the deep lock chambers now remaining. The short lengths between adjacent locks are connected to side pounds and there are several header pounds between groups of locks in the flight. Thus water released on the downward passage of a boat could be partially reused on later upward passages. The upper part of one lock chamber is widened in order to allow boats to pass.[1]

In all, the series of locks gave a rise of 168 ft in approximately half a mile of travel, giving an average rise of 12 ft per lock. Thomas Dadford, junior, was the Engineer for the canal and the Crumlin branch was completed in 1798.

The lock chambers have now been restored by British Waterways and at the head of the flight is a small Canal Museum and Interpretative Centre.

1. RUSSELL R. *Lost canals of England and Wales*. David and Charles, Newton Abbot, 1971, 115.

33. Ynys-y-Fro Reservoir

HEW 1200
ST 285 889

Near the upper end of the flight of 14 locks on the Monmouthshire Canal is the Ynys-y-Fro reservoir. The year 1846 saw the establishment of a private company to supply water to the town of Newport, which then had a population of 19 000, and the passing of a Parliamentary Act which authorized the company to construct a reservoir at Ynys-y-Fro and to lay pipes to convey water to the town. The Engineer was James Simpson.

The reservoir had a capacity of 71 million gallons and a maximum depth of 34 ft. It was contained by an earth dam with clay puddle core, having side slopes of about 1 in 2.5 and stone pitching on the submerged face.

In or about 1881 the company commissioned a further reservoir upstream and next to the original, the purpose apparently being to increase the storage capacity and to trap silt which was causing a discoloration of the water supply. The new dam was of similar construction to the earlier one; it now forms a causeway carrying a local road between the upper and lower reservoirs.

34. Newport Transporter Bridge

HEW 144
ST 318 862

Newport Transporter Bridge across the River Usk is about 1½ miles south of Newport town centre and was completed in 1906.[1] The only other similar bridges in Britain are at Middlesbrough[2] and Warrington.[3] Located in an industrial area, the bridge is a slender, attractive structure. The design was chosen in order to give adequate headroom for the type of shipping using the River Usk at the turn of the century: the bridge has a span, from centre to centre of the towers, of 645 ft and a clear headway from high water level of 177 ft.

The towers are of open lattice steel construction, each founded on four cast-iron cylinders. There are 16 suspension cables, four inside and four outside each of the stiffening girders, which are 16 ft deep and 26 ft 3 in. from centre to centre. The bottom booms are built-up plate girders, each lower flange carrying rails on which the travelling frame runs.

The platform car, which is 33 ft long by 40 ft wide, is hung from the travelling frame by 30 suspension ropes. The total weight of frame and car is just over 50 tons and they are moved by steel wire ropes wound on a drum worked by two electric motors, each of 35 brake horse power.

The Engineers were F. Arnodin and R. H. Haynes and the builders were Cleveland Bridge and Engineering Company of Darlington.

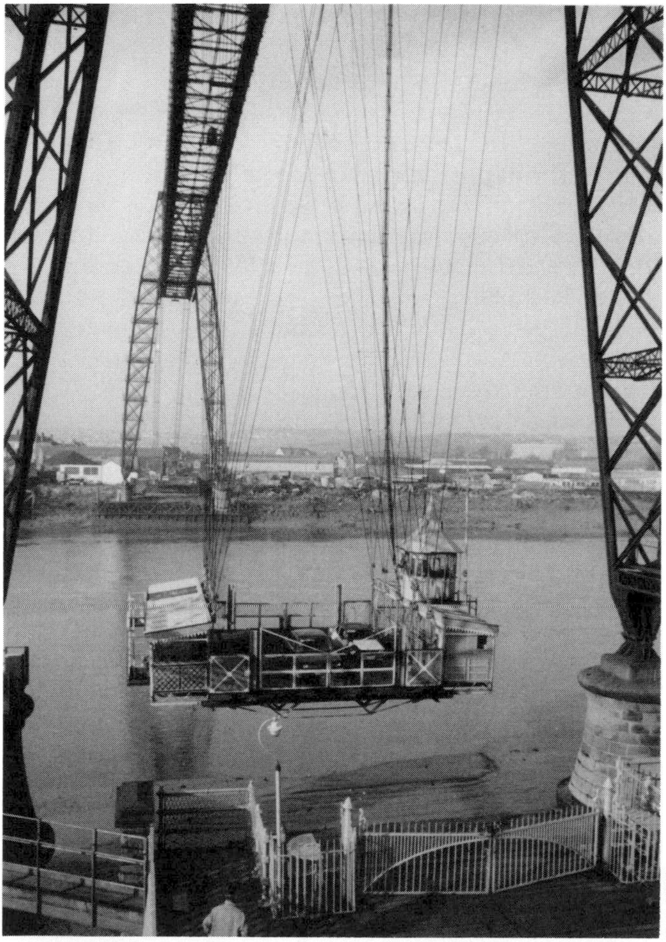

O. M. GIBBS

Newport Transporter Bridge

The bridge was closed to traffic in 1985[4] because of corrosion, but was reopened in 1995 following extensive restoration.

1. The transporter bridge at Newport. *Engineer*, London, 1906, **102**, 14, Sept., 263–265. (Abstract: *Min. Proc. Instn Civ. Engrs*, 1906–07, **167**, 405–406).

2. RENNISON R. W. *Civil engineering heritage: Northern England*. Thomas Telford, London, 1996, 90–91 (HEW 10).

3. Ibid. 267 (HEW 140).

4. DADSON J. Corrosion raises cable conundrum. *New Civil Engineer*, 1986, 20 Feb., 36–37.

35. Bigsweir Bridge

HEW 361
SO 539 051

Travellers on the A466 between Chepstow and Monmouth temporarily cross from Wales into England over this bridge, which spans the River Wye and blends harmoniously into its beautiful scenic setting. The bridge formed part of a toll road between St Arvans and Redbrook which was authorized in 1824, and was completed in 1828. The designer was Charles Hollis of London.

The elegant cast-iron arch of 164 ft span has four ribs of dumb-bell section in 16 segments with plain N spandrel bracing. It is supported on circular stone piers and is surmounted by a cast-iron balustrade. The arch segments, cast at Merthyr Tydfil, are particularly well made.

Around the mid-nineteenth century two masonry-arched flood spans were added at each end, making the overall length 321 ft. The mock string course cast on the main arch at deck level is continued in masonry over the faces of the flood spans.

Despite its age, and a carriageway only 12 ft wide, the bridge copes with the heavy tourist traffic along the Wye valley, with the aid of traffic signals and a 17 tonne weight restriction.

Bigsweir Bridge

R. CRAGG

R. CRAGG

Chepstow Bridge

36. Chepstow Bridge

HEW 145
ST 536 943

To the historically minded, mention of Chepstow brings to mind the medieval castle and Brunel's railway bridge (now replaced, but described in Chapter 5, No. 10). However, just upstream, and not far from Chepstow Castle, is an older and interesting five-span cast-iron bridge carrying the Chepstow to Tutshill road over the River Wye, and opened in 1816 to replace the timber bridge shown in J. M. W. Turner's painting of the castle.

The bridge was designed by J. U. Rastrick and cast by William Hazledine. The centre span is 112 ft, flanked by spans of 70 ft and 34 ft on each side. The spandrel fillings are of the radial grid pattern, not unlike those on some of Telford's bridges. The graduated spans, vertically curved road profile, parapet fence and decorative lamp standards combine to give the bridge its attractive appearance.

In spite of its age, the bridge has continued to carry traffic with the help of traffic signals, which limit the flow to one lane alternately in each direction. The centre span was strengthened in 1889 by the addition of steel ribs and the timber starlings have been improved by sheet steel piling, while extensive repairs were carried out in 1979–80.

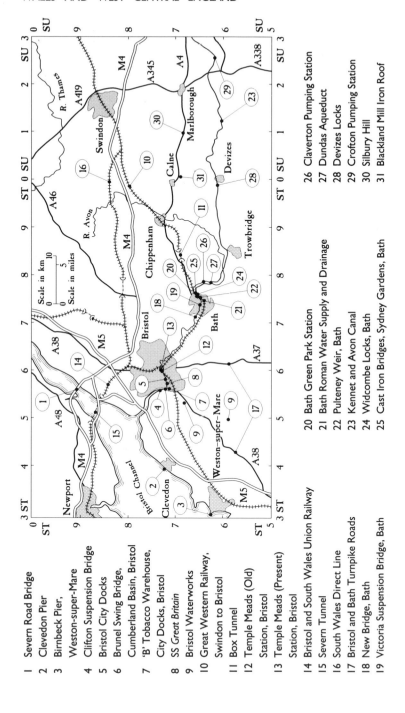

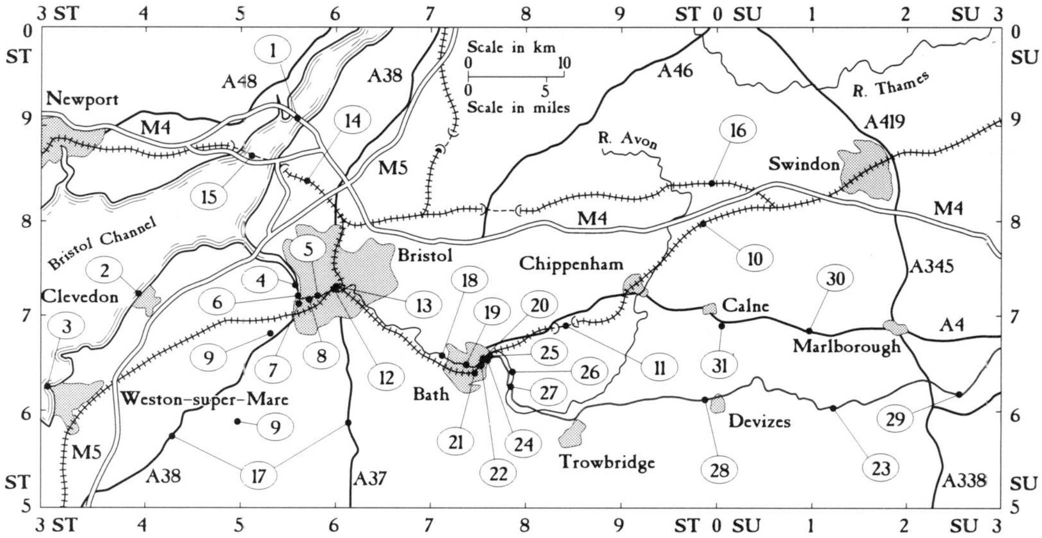

1 Severn Road Bridge
2 Clevedon Pier
3 Birnbeck Pier, Weston-super-Mare
4 Clifton Suspension Bridge
5 Bristol City Docks
6 Brunel Swing Bridge, Cumberland Basin, Bristol
7 'B' Tobacco Warehouse, City Docks, Bristol
8 SS *Great Britain*
9 Bristol Waterworks
10 Great Western Railway, Swindon to Bristol
11 Box Tunnel
12 Temple Meads (Old) Station, Bristol
13 Temple Meads (Present) Station, Bristol
14 Bristol and South Wales Union Railway
15 Severn Tunnel
16 South Wales Direct Line
17 Bristol and Bath Turnpike Roads
18 New Bridge, Bath
19 Victoria Suspension Bridge, Bath
20 Bath Green Park Station
21 Bath Roman Water Supply and Drainage
22 Pulteney Weir, Bath
23 Kennet and Avon Canal
24 Widcombe Locks, Bath
25 Cast Iron Bridges, Sydney Gardens, Bath
26 Claverton Pumping Station
27 Dundas Aqueduct
28 Devizes Locks
29 Crofton Pumping Station
30 Silbury Hill
31 Blackland Mill Iron Roof

4. The Bristol Area, Bath and North Wiltshire

The area covered by this chapter is dominated by the city and port of Bristol and its communications with London, the Midlands, South Wales and the South-West Peninsula. The city's industries grew up on locally produced raw materials and on imports of foreign goods, including sugar and tobacco from the Americas.

The Bristol coalfields extended in a wide arc from south-west to north-east, passing round to the east of the city, while further afield the Somerset coalfield, south-east of Bristol and south-west of Bath, was worked until 1975.

To the south the Mendips once provided lead, worked by the Romans, and zinc for brass making, and still produce limestone in large quantities. The Mendip area is also an important source of water supply. Bath is largely built of Bath stone, still mined to the east of that city. Iron ore was found locally and iron was also brought down the Severn from the Forest of Dean.

From the eighteenth century, expanding trade led to the construction of canals, the establishment of turnpike roads and at Bristol the building and improvement of the docks. In the nineteenth century came the railways, pioneered by I. K. Brunel's Great Western Railway to connect Bristol with London.

Today the area is traversed from east to west by the M4 motorway from London to South Wales which intersects at Almondsbury, on the outskirts of Bristol, with the M5 motorway from south-west England to the Midlands.

Many of the original local industries and the facilities which served them have disappeared or fallen into disuse, to be replaced by developments using modern technology. However, much still survives as a reminder of our civil engineering heritage and of the engineers themselves, names such as Jessop, Rennie and Brunel.

R. CRAGG

Severn Road Bridge

HEW 201
ST 560 901

1. Severn Road Bridge

The second Severn Crossing, completed in 1996, has relieved the original Severn Road Bridge of much of the traffic travelling along the M4 between England and South Wales. The older bridge, built between 1961 and 1966, may still be considered rather young to be regarded as an Historical Engineering Work, but it is important in terms of technical development. It may be compared with other outstanding British suspension bridges, listed in the table below.

Bridge	*Date*	*Span* (ft)
Union[1]	1820	361
Menai	1826	579
Clifton	1836–64	702
Tamar[2]	1961	1100
Forth[3]	1964	3300
Severn	1966	3240
Humber	1981	4626

The design of the Severn Bridge pioneered the arrangement of the hangers in a triangular pattern, while the deck is an integral streamlined box girder instead of the older type of trussed stiffening side girders with separate cross girders and decking.[4,5,6] The Humber Bridge and the Bosporus Bridge in Turkey follow the same pattern.

The main span is 3240 ft with 1000 ft side spans and the road is 120 ft above the water. The two steel towers are 400 ft high and 104 ft 6 in. wide and the main cables are 20 in. in diameter.

The substructure was designed by Mott, Hay and Anderson and built by John Howard; the superstructure by Freeman Fox and Partners and Associated British Bridge Builders (Arrol, Cleveland Bridge and Dorman Long).

A ten-span, 2023 ft long, steel viaduct across the Beachley Peninsula joins the Severn Bridge to the Wye Bridge, a cable-stayed cantilever bridge with a 770 ft main span and two 285 ft side spans.

Traffic volumes on the Severn and Wye bridges increased rapidly after their opening in 1966 and assessments of their structural capacity in 1982 showed that some elements had less than the required margin of safety against the worst possible loading. A comprehensive and innovative strengthening programme was carried out between 1985 and 1991.[7]

The Severn Bridge deck boxes were strengthened and the hangers replaced. In addition, internal systems of tubular steel columns, supporting plate girder grillages, were installed in the towers. The grillages were jacked up to relieve the towers and cable saddles of about 25 per cent of the loadings.

The original single stays of the Wye Bridge were replaced by a twin-stay system, which entailed strengthening and extending the towers.

Immediately downstream of the Severn Bridge is the longest overhead cable crossing[8] in Britain.

From the bridge, Oldbury Power Station[9] can be seen a few miles upstream on the east bank of the Severn. This was the first nuclear power station in Britain to have a concrete pressure vessel.

1. RENNISON R. W. *Civil engineering heritage: Northern England*. Thomas Telford, London, 1996, 9–10.

2. ANDERSON J. K. The Tamar Bridge. *Proc. Instn Civ. Engrs*, 1965, **31**, 337–360.

3. ANDERSON J. K. *et al.* Forth Road Bridge. *Proc. Instn Civ. Engrs*, 1965, **32**, 321–512.

4. ROBERTS G. Severn Bridge: design and contract arrangements. *Proc. Instn Civ. Engrs*, 1968, **41**, 1–48.

5. GOWRING G. and HARDIE A. Severn Bridge: foundations and substructure. *Proc. Instn Civ. Engrs*, 1968, **41**, 49–67.

6. HYATT K. E. Severn Bridge: fabrication and erection. *Proc. Instn Civ. Engrs*, 1968, **41**, 69–104.

7. FLINT A. R. Strengthening and refurbishment of the Severn Crossing. *Proc. Instn Civ. Engrs, Civ. Engng*, 1992, **92**, 57–65.

8. Aust Overhead Cable Crossing (HEW 1146) ST 556 891.

9. Oldbury Nuclear Power Station (HEW 1180) ST 606 943.

2. Clevedon Pier

HEW 430
ST 401 719

Further down the coast is Clevedon Pier, probably the most elegant of the seaside piers around Britain's coast.[1]

Eight 100 ft spans comprise a 16 ft 6 in. wide pier leading out to a pierhead. The four legs of each supporting trestle are made of pairs of Barlow rails, of 'top hat' section, redundant from the South Wales Railway, riveted foot to foot. Each leg is spigotted into a 2 ft diameter cast-iron screw pile. Near the top, one rail of each pair

Clevedon Pier

R. CRAGG

curves away under the deck to form, with the corresponding rail of the adjacent pair of legs, a semicircular transverse bracing assisted by a series of struts. Longitudinally, additional rails curved to the profile of a quarter ellipse, are riveted to the legs and to the underside of the main deck girders, with some open-spandrel infilling.

The simplicity and slenderness of the construction render the pier most pleasing visually. Designed by J. W. Grover[2] and R. Ward, with the ironwork fabricated by the Hamilton Ironworks, Liverpool, the pier was opened in 1869.[3] The pierhead, built as a replacement in 1893, is at an angle to suit the run of the tide, with enough depth of water to permit berthing of ships at all states of the tide.

In the heyday of pleasure steamers, the pier served many trips between Bristol and the resorts in the Bristol Channel as well as affording an amenity for holiday makers and anglers. Latterly, changing public tastes in leisure activities greatly reduced the use of the pier and consequently the revenue, with adverse effects on its maintenance. Unfortunately in 1970 two of the outer spans collapsed during a load test, leaving the pierhead inaccessible.

Since 1980 the pier has been in the hands of a Preservation Trust which, with initially only limited financial resources, set about a restoration programme. In 1984 grants totalling £1 million were made to the trust. The pier, with the exception of the pierhead, was dismantled in 1985 and carried by barge up the coast to Portishead dock for refurbishment. The restored components were floated back in 1988 and re-erected with the aid of a jack-up barge. Sailings from the pier recommenced in 1989, after a break of 20 years. With the help of a further substantial grant in 1995, including money from the National Lottery Fund, completion of the final stages of restoration, involving the pierhead and its elegant cast-iron pavilions, is now assured.

1. MALLORY K. *Clevedon Pier*. Redcliffe Press, Bristol, 1981.

2. GROVER J. W. Description of a wrought-iron pier at Clevedon, Somerset. *Min. Proc. Instn Civ. Engrs*, 1870–71, **32**, 130–136.

3. Opening of the pier at Clevedon, Somersetshire. *Ill. London News*, 1869, 10 Apr., 369–370.

R. CRAGG

Birnbeck Pier

HEW 434
ST 307 626

3. Birnbeck Pier, Weston-super-Mare

Birnbeck Pier is one of the few remaining piers designed by Eugenius Birch, who was responsible for no fewer than 14 seaside piers in Britain. It was built by the Isca Iron Company of Newport, Gwent, and was opened in the presence of Birch himself in June 1867.[1]

The pier is unusual in that the 'pierhead' is in fact the small Birnbeck Island, which was levelled and stepped to form a promenading area. Wrought iron girderwork 1150 ft long, with a 20 ft wide timber deck and supported on 15 clusters of eight piles, connects the island to the shore. A further 250 ft of pier structure, constructed subsequently, extends northward from the island to form a landing stage. The island also accommodates the house and launching slip for the Weston-super-Mare lifeboat.

The parapets incorporate ornamental lamps and decorative seating, with small projecting bays at intervals, which with the slender and economical ironwork of the structure, lend the pier an air of unmistakable Victorian charm.

Like Clevedon Pier, Birnbeck catered for the once-thriving pleasure steamer trade in the Bristol Channel, and, like Clevedon's, its maintenance suffered as a result of falling revenue. However, the National Piers Society designated 1996 as 'The Year of the Pier'. This, together with the evident success of the Clevedon Pier project, has given added impetus to long-term restoration proposals involving the Weston-super-Mare Pier Company, North Somerset District Council, Avon Industrial Buildings Trust and English Heritage, who consider that the pier's neglect is a matter for national concern.

1. The new pier at Weston-super-Mare. *Ill.London News*, 1867, 15 June, 600–601.

4. Clifton Suspension Bridge

HEW 129
ST 565 731

The Clifton Gorge, where the River Avon cuts through the high ground of Clifton Down, has always been an impediment to travel in Bristol. In 1753 one William Vick left £1000 to be invested until £10 000 had accumulated, when the money was to be applied to realizing his dream of a bridge across the gorge.

In 1829 the trustees advertised a design competition, which, after a fierce and controversial contest, was won by I. K. Brunel, then aged 25. He was appointed Engineer[1,2] and work began in 1836, but by 1842 the money

SS *Great Britain* passing under Clifton Suspension Bridge

had run out when only the towers and anchor tunnels had been constructed. The chains, which had already been made by the Copperhouse Foundry, Hayle, were sold to form part of Brunel's Royal Albert Bridge at Saltash.[3]

After Brunel's death in 1859, a number of members of the Institution of Civil Engineers decided to complete the bridge as a memorial to him. Work began in 1860 and was completed in 1864 to amended designs by John Hawkshaw and W. H. Barlow.[4]

At 702 ft 3 in. the bridge had, for 97 years, the longest span in Britain. Its 31 ft wooden deck is 240 ft above river level. The suspension chains are made up of 7 in. x 1 in. wrought iron links, 24 ft long. Fittingly, some of them came from the original Hungerford footbridge across the Thames in London, the only other suspension bridge designed by Brunel; completed in 1845, this was demolished to make way for the Charing Cross Railway Bridge.

1. PORTER GOFF R. F. D. Brunel and the design of the Clifton Suspension Bridge. *Proc. Instn Civ. Engrs*, 1974, **56**, 303–321.

2. PUGSLEY Sir A. (ed.) *The works of Isambard Kingdom Brunel*. Institution of Civil Engineers and University of Bristol, 1976, 51–68.

Bristol City Docks circa 1977. (Redrawn from Port of Bristol Authority map)

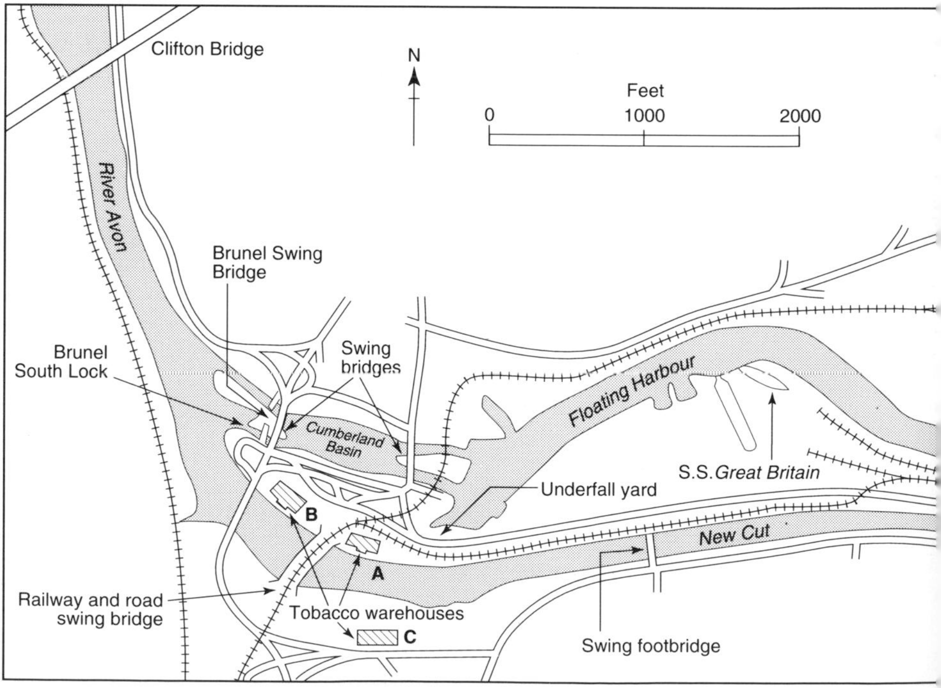

3. OTTER R. A. *Civil engineering heritage: Southern England.* Thomas Telford, London, 1994, 41 (HEW 29).

4. BARLOW W. H. Description of the Bristol Suspension Bridge. *Min. Proc. Instn Civ. Engrs*, 1866–67, **26**, 243–257.

5. Bristol City Docks

HEW 861
ST 567 723 to
ST 616 726

Within sight of the Clifton Suspension Bridge is the entrance to the City Docks which traverse Bristol from west to east.

The history of the docks begins in 1247 with the construction of the Broad Quay in the River Frome. From then on only relatively minor works were carried out until the increasing size and number of ships in the eighteenth century led to the creation, between 1804 and 1809, of the 'Floating Harbour', one of the greatest civil engineering undertakings of its time.

Designed by William Jessop,[1] this dock system comprised parts of the courses of the Rivers Avon and Frome, made possible by digging a new channel for the Avon (the 2 mile long 'New Cut') from Totterdown to Rownham, where an 'overfall' dam was constructed to maintain constant high water level in the Floating Harbour. At

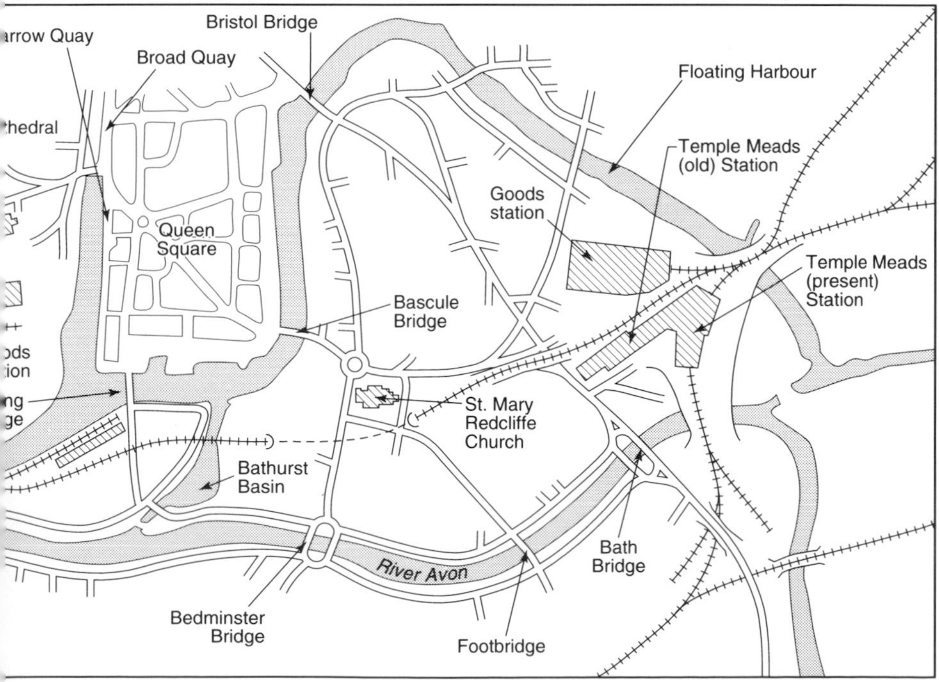

Rownham there were two entrance locks from the Avon into an entrance basin ('Cumberland Basin'), which was joined to the harbour by a Junction Lock on its south side. Water was fed to the harbour by a feeder canal from a dam across the Avon at Netham, 1 mile upstream of Totterdown. Barge locks, long since closed, communicated with the New Cut at Totterdown and Bathhurst Basin.

In 1832, on the advice of I. K. Brunel, four sluices were constructed through the overfall dam at Rownham—three for regulating water flow and one for scouring silt. This 'Underfall Dam' is now a scheduled ancient monument. A steam drag boat was also built at Brunel's suggestion for moving silt from the dock walls. In 1843 an improved drag boat to Brunel's design was built and this continued in use until 1961.

Brunel was the Engineer for a new and larger South Entrance Lock,[2] 262 ft long and 52 ft wide. Completed in 1849, the lock had the first ever wrought iron buoyant gate. This was a single leaf, hinged on the south side and supported on roller wheels.[3] The lock is now disused and the gate has been removed but much of the original structure can still be seen at low water.

In 1873 Thomas Howard constructed the present entrance lock to the north of Jessop's North Entrance Lock, which was sealed off. A new Junction Lock was also constructed on the north side of the Cumberland Basin. From then until the 1960s, various related minor but important alterations and improvements were made to the system.

The docks remain a statutory port but use is now almost entirely confined to leisure activities; the last commercial trader ceased operation in 1993. Bristol City Council, which acquired the docks in 1848, is developing the area as a public amenity with strong emphasis on preserving the many surviving examples of its engineering past.

1. HADFIELD C. and SKEMPTON A. W. *William Jessop, engineer.* David and Charles, Newton Abbot, 1979, 222–242.

2. South Entrance, Bristol Docks (HEW 1173) ST 567 723.

3. BUCHANAN R. A. I. K. Brunel and the Port of Bristol. *Trans. Newcomen Soc.*, 1969–70, **42**, 41–56.

R. CRAGG

6. Brunel Swing Bridge, Cumberland Basin, Bristol

Brunel Swing Bridge

HEW 926
ST 568 724

This interesting relic of Brunel's early girder work is still to be seen, swung parallel to the present entrance lock in the shadow of the large 1965 swing bridge over the lock. Originally 120 ft 9 in. long, it has wrought iron main girders with tubular flanges, the top circular, the bottom triangular. Twin iron tie bars run from end to end within the top flange.

The history is a little complex. The bridge was built in 1849 to span Brunel's South Entrance Lock.[1] Thomas Howard's North Entrance Lock, opened in 1873, was intended to supersede Brunel's lock; as a result the bridge was shortened by 10 ft and re-sited in its present position at the new entrance lock. However, pressure from steamship owners resulted in the South Entrance Lock being kept operational into the 1880s and consequently a replica of Brunel's swing bridge was built to span it. Now fixed, this later bridge still provides vehicular access to the island between the two locks.

1. South Entrance, Bristol Docks (HEW 1173) ST 567 723.

R. CRAGG

'B' Tobacco Warehouse, with 'A' Warehouse in the background

HEW 1011
ST 568 722

7. 'B' Tobacco Warehouse, City Docks, Bristol

'B' Tobacco Warehouse is one of three imposing multi-storey brick-clad warehouses, almost identical in appearance, near the Cumberland Basin entrance to Bristol City Docks.

'A' Tobacco Warehouse, on the north bank of the Avon (ST 570 721), was built between 1903 and 1905 as a steel-framed structure with floors of steel joists and concrete jack arches, founded on concrete footings as much as 50 ft below ground.

When the construction of No. 1 Granary at Avonmouth Docks[1] went ahead in reinforced concrete, it was decided to use that material for 'B' Warehouse. 'B' Warehouse was built between 1906 and 1908 by William Cowlin and Sons of Bristol, who also built Warehouses 'A' and 'C'; the latter (ST 569 720), built on the south bank of the river in 1915, was also in reinforced concrete. The designer, Edmond Coignet, was the son of François Coignet, a pioneer of structural concrete in France.

All three buildings are 214 ft long, 104 ft wide and 93 ft 9 in. high to the top of the parapet. The ground floors are

15 ft 6 in. high and the remaining eight floors 8 ft 6 in. high, to suit the stacking of cylindrical 'tierces' of leaf tobacco two high. The roofs are of north light steel trusses clad in slates and patent glazing. Tobacco Warehouse 'B' is founded on reinforced concrete piles, whereas Warehouse 'C' was set direct on rock near the surface.

With the changes in the handling of leaf tobacco in the 1960s, the warehouses ceased to fulfil their original function but are still in use, 'A' and 'C' for general warehousing. Warehouse 'B', now owned by Bristol City Council, houses the Bristol Archive and Record Office and a Community Project Centre.

1. No. 1 Granary, Avonmouth (HEW 1012) ST 513 785.

8. SS *Great Britain*

HEW 247
ST 578 723

Further up the Docks, Brunel's *Great Britain* lies in the former Great Western Dry Dock, which was built for her construction. She is being restored there to her original form as a memorial to one of the most historic ships ever built and to her great designer, whose versatility she exemplifies.[1,2]

SS *Great Britain* was the largest ship of the time, with a displacement of 3675 tons, and the first with iron hull and screw propulsion. She carried 252 passengers, 130 crew, 1200 tons of cargo and 1200 tons of coal. Her engines, using steam at 5 lb/sq. in., developed 1600 h.p. at 18 rev./min. The six-bladed propeller, 15 ft 6 in. in diameter, was driven at 54 rev./min. through four sets of chain drives. On trials the ship achieved a speed of 11 knots under steam alone. Sails, on six masts, were provided to assist propulsion when the wind was favourable.

Launched on 19 July 1843 by the Prince Consort, she made four voyages between Liverpool and New York before running aground in Dundrum Bay, County Down, in 1846. After an ingenious salvage operation by Brunel's nautical adviser, Captain Claxton, in August 1847, she was sold to new owners in 1850. They changed the propeller to three-bladed, installed new boilers at 10 lb/sq.in. and increased the passenger accommodation to 730.

SS *Great Britain*

In 1852 the ship began sailing between Liverpool and Australia, and by 1876 she had made 32 such voyages plus two to New York. She had also been requisitioned as a troop carrier in the Crimean War, 1853–56 (in which Brunel was also concerned as the designer of prefabricated timber hospital buildings) and the Indian Mutiny of 1857–58. In 1882 the ship was sold and converted for carrying coal under sail between South Wales and San Francisco, but in May 1886 she was damaged off Cape Horn and put into Port Stanley, in the Falkland Islands. There she served as a coal and timber store ship until in 1937 she was deliberately holed and beached.

In 1970 the SS *Great Britain* Project, in a remarkable salvage operation, raised the ship onto a pontoon on which she was towed back to Bristol. Released from the pontoon in the Avonmouth dry dock, she was towed up the Avon and on 19 July 1970 was berthed in the Great Western dry dock.

Since 1970 a steady programme of restoration of the exterior and interior of the great ship has been carried out. The propeller has been replaced with a replica of Brunel's original six-bladed design, together with the balanced rudder. A replica of the original engine is being built and will be installed in the engine room, where it will be turned by electricity. The interior cabins are also being progressively restored, in particular the lower dining saloon which has been recreated in its original style with ornate columns.

1. CORLETT E. G. B. The steamship Great Britain. *Trans. R. Soc. Nav. Archit.*, 1971, 113, 411–437.

2. CORLETT E. G. B. *The iron ship*. Moonraker Press, Bradford-on-Avon, 1975.

9. Bristol Waterworks

HEW 1242

The population of Britain nearly doubled during the second half of the Industrial Revolution. Towns grew and became congested. Water supply and sewage disposal, never good, became totally inadequate. The 1842 report *The Sanitary Conditions of the Labouring Classes in Great Britain* triggered off a great burst of sewerage schemes which in turn increased the demand for water, for domestic use as well as for industry.[1] Drainage into streams and rivers led to pollution and new sources of supply had to be sought. Many of the new developments took place in the same period as the great expansion of railways and, like them, placed considerable demands on the civil engineering resources of the nation.

In 1844 Bristol was described as having the worst water supply of any large town. In 1846, the Bristol Waterworks Company was incorporated and appointed James Simpson as Consulting Engineer.[2] He recommended bringing in the principal supply from springs in the Mendips, south of the city; in these hills, advantage could be taken of a 50 square mile catchment area at a height of 400 ft, with an annual rainfall of 41 in.[3]

The works involved collecting spring water from the Mendips in the vicinity of Chewton Mendip and conveying it in an aqueduct to a reservoir at Barrow (ST 535 673), on an average gradient of 5 ft in a mile. There are 4 miles of 30 in. diameter pipe and 4 miles of tunnel. On the route of the tunnel there are remarkable crossings of valleys at Winford, Leigh and Harptree by wrought iron riveted tubes, the first two 825 ft long and the third 350 ft. The tubes are egg-shaped, 4 ft 7½ in. high and 3 ft 6 in. wide, supported on cast-iron saddles on the tops of masonry piers at 50 ft centres; the piers are as much as 60 ft high, and cast-iron ball bearings allow longitudinal movement. The aqueduct is still in service today and forms an integral part of the Mendip supply system.[4]

From Barrow the water was piped to service reservoirs at Clifton, Durdham Down and Bedminster, then on the outskirts of the city. The work was completed in 1851 and one of the original mains is still in service. After 130 years, this 20 in. diameter cast-iron main was found to be badly corroded on the outside, but in fair condition internally. In 1983–84 2½ miles of the main were lined with a continuously welded polyethylene pipe pulled through from excavations made at 100 yd intervals. This was the first application in the area of this technique, which avoids the wholesale digging up of pipelines for renewal.

The initial works were designed for 4 million gallons per day. In 1866 a further storage reservoir was completed at Barrow and a third was added in the period

1897–1900. Between 1867 and 1870 a further 6 million gallons per day were pumped from springs at Chelvey[5] to Barrow. Simpson's 60 h.p. pumping engines at Chelvey were scrapped in 1937.

In the period 1898–1901 a 700 yd long earth dam was built across the Yeo valley near Blagdon (ST 504 600) which provided a 470 acre reservoir with 1700 million gallons capacity. In 1930 springs in the Cheddar Gorge were tapped for pumping to Barrow and in 1937 Cheddar reservoir (ST 442 538) was completed to store the excess water for use during dry periods.

In 1956 HM Queen Elizabeth II opened Chew Valley Lake, at the time the largest man-made lake in Britain, and still one of the six largest. An earth dam (ST 571 616) nearly ¼ mile long, with a concrete filled cut-off trench into the underlying marl, impounds 4500 million gallons in a lake 2¼ miles long and 1½ miles wide. This serves not only Bristol but also as far afield as Bath, Radstock and south-west Gloucestershire.

Blagdon and Chew Valley Lake, with the Barrow reservoirs, are important trout fisheries, their trout hatchery having a world wide reputation. Cheddar and Chew offer small boat sailing.

In 1965 the Company began pumping Severn water from the Gloucester and Sharpness Canal at Purton (SO 695 038), initially to treatment works at Littleton (ST 604 894). Purton treatment works (SO 698 036) was completed in 1973 and rebuilt and expanded in 1994–95.

1. Green J. Account of recent improvements in the drainage and sewerage of Bristol. *Min. Proc. Instn Civ. Engrs*, 1848, **7**, 76–84.

2. Binnie G. M. *Early Victorian water engineers*. Thomas Telford, London, 1981, 82–91.

3. Coysh A. W. *et al. The Mendips*. Robert Hale, London, 1977, 196–201.

4. Simpson J. Discussion: Bridge-aqueduct at Roquefavour, by G. Rennie. *Min. Proc. Instn Civ. Engrs*, 1854–55, **14**, 218–223.

5. Chelvey Pumping Station (HEW 1539) ST 473 679.

10. Great Western Railway, Swindon to Bristol

HEW 1071
SU 149 856 to ST 597 724

The Great Western Railway (GWR) between Bristol and London, authorized under an Act of 31 August 1835, was

I. K. Brunel's first venture into the field of railway engineering, in which he was to become a master.

From London to the summit of the line at Swindon and a little way beyond, gradients were easy and few major engineering works were necessary other than at river crossings, but over the next 41 miles west of Swindon the terrain was more formidable and numerous major works were required. This suggested that higher speeds might be attained on the London side, using locomotives with larger diameter driving wheels than those used westwards.

A pause to change engines and allow the passengers time for refreshment was not unreasonable in 1840, when the new-fangled railway was replacing horse-drawn stage coaches which operated at about one-quarter of the speed and with a similar fraction of the distance between stops.

At the outset, in 1840, on the advice of Daniel Gooch, Brunel's young Locomotive Superintendent appointed at the age of 25, the GWR established a locomotive depot or running shed at Swindon to cater for 48 locomotives under cover and to provide repair facilities for a further 54. This developed into Swindon Works—one of the largest railway establishments in the world for the construction and repair of locomotives, carriages and wagons. It eventually occupied about 320 acres, of which over 70 acres were roofed. At its peak there was a staff of 12 000 and a technical and manufacturing capability almost unrivalled in its day.

Between Swindon and Chippenham, deep cuttings and high embankments were required over poor ground.

The 13 miles from Chippenham to Bath were the hardest of the line. Starting with the viaduct at Chippenham[1] there were long embankments and deep cuttings in rock; then Box Tunnel (see below), followed by Middle Hill Tunnel[2] and a bridge across the River Avon at Bathford (ST 785 671). At Bath the Avon was crossed just east of the station by an 88 ft span stone arch (rebuilt 1926–27). Two 80 ft laminated timber arches,[3] replaced by an iron girder bridge in 1879, crossed the river west of the station. On both sides of the Avon the bridges are approached by long stone viaducts.

Isambard Kingdom Brunel, Engineer of the Great Western Railway

Between Keynsham and Bristol a number of short tunnels were needed to take the railway along the steep hillside above the River Avon. Outside the Bristol terminus, three bridges were needed to cross waterways. Two have been replaced by modern steel structures. The survivor, Brunel's 'Gothic' masonry arch[4] across the Avon (ST 614 724), is itself masked from view by modern bridges on each side.[5]

The line was opened throughout from London to Bristol on 30 June 1841, following completion of the Box Tunnel.

Brunel had adopted the broad gauge of 7 ft ¼ in. on the grounds that this would permit larger locomotives, higher speeds and great comfort and safety for the passengers. But the 'standard' gauge of 4 ft 8½ in. had been firmly established on the other railways which had spread since the Stockton and Darlington Railway[6] was opened in 1825, and in 1846 the Gauge Act came down firmly on the side of the standard gauge for further railways.

Nevertheless, the GWR broad gauge penetrated to the extremity of the south-western peninsula and into South Wales, although, quite early on, a third rail was added to various sections of track in order to permit through running with other lines. London to Bristol had mixed gauge by 1875. In 1869, by which time there were 1500 miles of broad gauge track, the Company had conceded that the broad gauge had lost the contest and embarked on a programme of conversion which was not completed until 1892.

Relics connected with I. K. Brunel may be seen at the GWR Museum at Swindon.

1. Chippenham Viaduct (HEW 380) ST 912 731.

2. Middle Hill Tunnel, Box (HEW 381) ST 820 687.

3. Pugsley Sir A. (ed.) *The works of Isambard Kingdom Brunel*. Institution of Civil Engineers and University of Bristol, London and Bristol, 1976, 121–122.

4. Gothic Arch, Temple Meads (HEW 1108) ST 614 724.

5. Brunel I. K. Calculation book. 1837 (Library, University of Bristol).

6. Rennison R. W. *Civil engineering heritage: Northern England*. Thomas Telford, London, 1996, 83–84 (HEW 85).

HEW 236
ST 830 689 to
ST 857 694

11. Box Tunnel

Box Tunnel, some 5 miles east of Bath, is the most important of the substantial works necessary to bring Brunel's Great Western Railway down to Bristol from the summit level at Swindon.

One mile and 1452 yd long, with a downward gradient of 1 in 100 towards Bath, it was then, by nearly 800 yd, the longest railway tunnel so far built. It is 30 ft wide, which accommodated two of Brunel's broad gauge tracks.

Work began in 1836 with the sinking of seven shafts,[1] the deepest 290 ft, up which the excavated material was raised by horses at the surface operating hoisting drums.

R. CRAGG

Box Tunnel—west portal

Two contracts were let for the driving of the tunnel. In the contract for the westernmost half mile, which was in rock and unlined, excavation was by blasting with gunpowder, of which a ton was used each week. Considerable ingress of water was dealt with by two 50 h.p. pumps.

In the remainder of the tunnel, in softer strata, excavation was by pick and shovel, using techniques which had been developed over many years in the construction of canal tunnels, and the tunnel was lined with brick. Brunel had, of course, already gained experience in this class of work in helping his father on the Thames Tunnel, opened in 1843.

Work on the Box Tunnel was completed in 1841. At its peak, 4000 men and 300 horses were employed. The deaths of 100 men during the progress of the works underline the hazards of construction.

The monumental west portal is a notable feature.

1. BRUNEL I. K. *Private letter book*. 10 Oct. 1836 (Library, University of Bristol).

12. Temple Meads (Old) Station, Bristol

HEW 224
ST 596 724

The original terminus of the Great Western Railway (GWR) was opened in 1841. Behind the neo-Tudor frontage of the offices and GWR boardroom, engine and train sheds were supported on brick-arched vaults. Above three of the five running tracks in the engine shed, cast-iron columns carried the drawing office where GWR schemes were designed. Beyond the engine shed was a 220 ft long train shed covered by the station's most notable feature, its timber roof, the centre span of which, at 74 ft, is wider than the roof of Westminster Hall.[1]

The roof is basically a series of cantilevers, supported on lines of cast-iron columns just inside the platform edges, meeting at the ridge, and tied down at their rear ends to the outside walls of the building. Brunel evidently felt that such simple construction on its own was not worthy of the western terminus of his great undertaking, the Great Western Railway, and added further embellishments in the shape of curved brackets, with lattice infilling, at the springings of the cantilevers, together with ornamental timber pendants to give the appearance of a hammerbeam roof.[2]

Brunel roof, Temple Meads Old Station

R. CRAGG

The train shed was extended in the 1870s to almost twice its original length.

No longer in use, the old station served for a number of years as a covered car park. Brunel's train shed has been restored by a Trust and now houses a museum.

1. BRUNEL I. K. Discussion: The construction of the collar roof with arched trusses of bent timber at Horsley Park, by the Earl of Lovelace. *Min. Proc. Instn Civ. Engrs*, 1849, **8**, 285–286.

2. TOTTERDELL J. A peculiar form of construction. *J. Bristol & Somerset Soc. Archit.*, 1960, **5**, Mar., 111–112.

13. Temple Meads (Present) Station, Bristol

HEW 436
ST 598 725

The main Great Western Railway line from London joined the Bristol and Gloucester Railway on an east–west alignment before swinging south-west over the floating harbour into the original Brunel terminus. Adjacent to this, but at right angles, was the terminus of the Bristol and Exeter Railway,[1] whose line set off southwards across the New Cut and whose 1854 Jacobean-style offices can be seen on the south-east side of the present station approach.

By 1865 enlargement and alterations became necessary and in 1876 a new station roof was built over a 970 ft radius curve, linking all routes as part of the Great Western and Midland Joint Station.

A rather similar modernization took place at York at much the same time and it is interesting to compare the two roof designs. Whereas York[2] has four spans of wrought iron plate girders without ties, Bristol has one 125 ft span consisting of 26 lattice ribs at 18 ft 9 in. centres, trussed with 3½ in. and 3 in. diameter main ties and 4 in. diameter tube struts.

The station façade, at the head of a wide approach ramp, is in an ornate Gothic style.

1. OTTER R. A. *Civil engineering heritage: Southern England*. Thomas Telford, London, 1994, 105–107 (HEW 1072).

2. RENNISON R. W. *Civil engineering heritage: Northern England*. Thomas Telford, London, 1996, 148–149 (HEW 239).

14. Bristol and South Wales Union Railway

HEW 1033
ST 609 726 to
ST 504 882

The Bristol and South Wales Union Railway (B&SWUR) represented one of the steps in reducing the distance by rail to South Wales.[1] Opened in 1863[2] for passengers only, 11 miles of broad gauge track shortened the distance from Bristol to Cardiff from 94 miles to 38 as compared with the previous route via Gloucester. The Engineer was R. P. Brereton, I. K. Brunel's principal assistant.

The line led from a junction with the Great Western Railway east of Temple Meads, via a deep cutting at Horfield (ST 607 776) and 1245 yd of tunnel at Almondsbury (ST 598 824) to New Passage (ST 543 863) on the east bank of the Severn. Here the trains ran along a timber jetty 1635 ft long. A steam ferry made the 2 mile crossing to a similar jetty, 708 ft long, at Black Rock (ST 514 881) whence about a mile of railway connected with the Gloucester to Newport line at Portskewett. The jetty heads incorporated stairs and ramps and floating pontoons. Marc Brunel, the father of Isambard, is reputed to have been the first to suggest the idea of floating pontoons as landing stages on a visit to Liverpool in 1826. When the Severn Tunnel was opened in 1886 the ferry was abandoned but much of the railway work on the Bristol side was incorporated in the new route.

At Black Rock a few pile stumps remain and just back from the foreshore is a masonry arch carrying a footpath over the cutting that led to the jetty.

At New Passage the line of the jetty can just be traced in the foreshore mud and a short length of the masonry abutment is built into the sea wall known as the Binn Wall.[3] This wall extends from Severn Beach to just north of New Passage, and forms part of the sea defences of the low-lying land between Avonmouth and Aust. Originally dating from at latest the first half of the seventeenth century, the wall has been progressively raised, widened and reconstructed.[4]

Immediately beyond the Binn Wall is a natural bank of gravel. In 1823 J. L. McAdam, Surveyor to the Bristol Turnpike Trust, was fined by the Wall Commissioners for removing gravel for use on his roads and later the B&SWUR had to be restrained from using this material.

1. Barrie D. S. M. The Bristol and South Wales Union Railway. *Rly Mag.*, 1936, **79**, Dec., 423–427.

2. The Bristol and South Wales Union Railway. *Ill. London News*, 1863, 5 Sept., 245.

3. Binn Wall (HEW 1207) ST 539 860 to ST 545 865.

4. Cole S. D. *The sea walls of the Severn*. Privately printed, Bristol, 1912.

15. Severn Tunnel

HEW 232
ST 480 876 to ST 545 854

The construction of the Severn Tunnel, the longest railway tunnel in Great Britain (4 miles 624 yd), further shortened the route between London and South Wales and is the story of a battle against nature in the shape of groundwater and the tides, well described in the classic book by T. A. Walker the contractor.[1] The Engineer was Sir John Hawkshaw.

The tunnel crosses the Severn between Pilning in South Gloucestershire and Sudbrook in South Wales, where the estuary is 2¼ miles wide and the range of the tide can be 50 ft. Although a great deal of the rocky bed is uncovered at low tide, the main channel, known as the 'Shoots', is 80 ft deep below general bed level.

The Great Western Railway began the work in 1873 with the sinking of a shaft at Sudbrook and the driving of a drainage heading towards the river. Four and a half years later only the shaft and 1600 yd of heading had been built. In 1877 contracts were let for additional shafts on both sides and headings on the line of the tunnel. In October 1879, when the headings had nearly joined up, ingress of fresh water from an underground river, the Great Spring, inundated the original Sudbrook workings. A contract was then let to Walker for the completion of the whole works, but it was not until January 1881 that the spring was contained within a 300 yd section of the tunnel by two headwalls; this enabled the rest of the works to be dewatered.

In April 1881 water broke into the Gloucestershire workings from a hole in the river bed. This hole was sealed with clay in bags and with concrete. In October 1883 the Great Spring broke in again and a week later a spring tide flooded the long deep cuttings on both sides. A heading from the Sudbrook shaft was driven to intercept the Great Spring and with the water so diverted, the final 300 yd of tunnel within which the spring had previously been walled could be completed.

Two 50 in. Bull engines, made by Harvey of Hayle in Cornwall, and installed for dealing with the flooding, still survive, one in the Science Museum, London and one stored by the National Museum of Wales, Cardiff.

The brick lining, 2 ft 3 in. to 3 ft thick, was completed for the full length of the tunnel in April 1885 and in September a special train took a party through the tunnel which included Sir Daniel Gooch, at that time the Chairman of the Company.

In 1886 a shaft was sunk beside the tunnel and six pumps with 70 in. beam engines, also by Harvey of Hayle, were installed to deal with the 20 million gallons of water per day, mainly from the Great Spring, at the Sudbrook pumping station. Three of these engines remained in use until the station was electrified in 1961.

The tunnel was opened for goods traffic on 1 September 1886, and for Bristol to Cardiff passenger traffic three months later. The first London to South Wales passenger train passed through the tunnel on 1 July 1887.

From the Gloucestershire side there is a falling gradient of 1 in 90 and from the Welsh side of 1 in 100, to the deepest section under the Shoots.

In 1995 Railtrack, the tunnel owners, began an extensive programme of pump replacement and the modernization of the control system. Sudbrook pumping station now supplies water from the Great Spring to a nearby paper mill, a brewery and the Ministry of Defence.

1. WALKER T. A. *The Severn Tunnel: its construction and difficulties, 1872–1887*. Bentley, London, 1891; Kingsmead Reprints, Bath, 1969.

16. South Wales Direct Line

HEW 1073
SU 067 818 to
ST 611 790 and
ST 612 804

This railway was opened by the Great Western Railway on 1 May 1903 from their main London to Bristol line at Wootton Bassett to junctions with the Bristol to Severn Tunnel line at Filton and Patchway.

The new line shortened the distance between London

and South Wales by 10 miles and relieved the tracks east of Bristol from the overloading which followed the opening of the Severn Tunnel—among other traffic great quantities of Welsh steam coal were being carried for the Navy at Portsmouth and the ocean liners at Southampton.

The crossing of the southern end of the Cotswold Hills entailed some substantial engineering works, the most notable of which are the Sodbury Tunnel, 2 miles 924 yd long (ST 793 812 to ST 752 815) and three massive brick viaducts at Winterbourne, of which the largest at ST 658 800 has eleven spans and is 265 yd long and 90 ft high.

17. Bristol and Bath Turnpike Roads

'Our shops, our horses' legs, our boots, our hearts have all benefited by the introduction of Macadam.'

(Charles Dickens)[1]

HEW 1061
ST 58, ST 98, ST 45 and ST 95

The system by which the construction and maintenance of roads was financed by the payment of tolls at 'turnpike' gates dates from the middle of the seventeenth century, but it was not until the beginning of the next century that the idea really took hold. Demands for speedier communications to serve government, the growing industries and the patrons of the newly popular spas and seaside resorts led to a great expansion which continued until the advent of railways in the 1830s.

Individual Acts of Parliament gave powers to trustees to raise money by public subscription for the initial improvement of highways and for the installation of gates, toll-houses, weighbridges, milestones and signposts, and to charge users tolls for maintenance and the servicing of the debt. The trusts were served by a clerk, an accountant and one or more surveyors whose field of duty covered the area which could be ridden on horseback in a day.

The Acts for the Bristol (1727) and Bath (1707) Trusts were followed by amending Acts which eventually led to the Bristol Trust covering 178 miles of road—thus making it the largest trust outside London—and the Bath Trust, 73 miles.

Most of the major roads out of these cities today follow

the routes of the turnpike roads, although inevitably, with improvements and the provision of bypasses over the years, there have been local deviations. Examples of such roads are the A38 (Gloucester and Bridgwater) and the A37 (Wells and Shepton Mallet) out of Bristol; the A4 (The Bath Road) towards London; and the A39 (Wells) and A367 (Shepton Mallet) out of Bath.

The Bristol and Bath Trusts are notable not only for having been among the earliest, but also for their association with the pioneering work of John Loudon McAdam, who was appointed General Surveyor to the Bristol Trust in 1816 and the Bath Trust in 1826, holding the former post until 1825 and the latter until his death in 1836.[2]

McAdam made his mark as much in the field of management as in that of construction.[3] He insisted on surveyors being responsible to the general surveyor instead of, as hitherto, individually to the trustees, and called for fortnightly reports on work done, with returns of labour, materials, transport and all costs, including contract work.

He was more a road improver than a road builder. On existing roads he had the top foot of stone taken up, where as much existed, broken into pieces not exceeding 6 oz in weight and then relaid in two 6 in. layers. For new roads he used the same grade of broken stone, in a single layer, sometimes as little as 6 in. thick. In this he differed from Telford, who insisted on a substantial foundation of coarse pitched stone, topped by a McAdam-type surface layer.

Although McAdam's system was much cheaper than Telford's, the quality of his roads permitted, as did Telford's, a doubling of speeds to as much as 12 miles per hour. This in turn was made possible only by a corresponding advance in the design and construction of coaches.

Horses' hooves and iron tyred wheels ground the surface stones into dust which served as a binder, but this happy situation ended when motor vehicles came on the scene. Their rubber tyres and higher speeds sucked out the finer particles, broke up the road surface and produced clouds of dust. Dressing the surface with tar and sand and eventually the use of pre-coated stone was the remedy—hence 'tarred macadam' and eventually 'tarmac'.

In 1888 the responsibility for public highways was transferred by statute to the local authorities. The turnpike system, with its gates and toll-houses, lapsed completely; but there are still many relics remaining. In 1967 the Bristol Industrial Archaeological Society deposited with the City Museum a comprehensive index of toll-houses, milestones, boundary posts, finger posts etc. resulting from an inspection of all roads within 40 miles of Bristol. Somerset Industrial Archaeology Society has recorded roadside features along those lengths of the Bristol and Bath turnpike roads which lie in Somerset.[4,5]

There is an interesting toll-house, fronted by cast-iron columns, at Ashton Gate, Bristol at the junction of North Street and Ashton Road (ST 572 718). This was originally at the beginning of the Bridgwater road but now, as the result of urban development, is virtually in a side street. At ST 597 636 on the B3130, which branches off the A37, a charming little two-storey toll-house stands on an island at a road junction. At Red Post (ST 698 571) on the A367 at Peasedown St John is one of the Bath toll-houses, and diagonally over the crossroad there, a cast-iron road marker dated 1827 is set at the foot of the wall of the Red Post Inn.

1. Webb S. and Webb B. *English local government: the story of the King's highway*. Longmans, 1913; reprinted by Frank Cass, London, 1963, 184.

2. McAdam J. L. *Narrative of affairs of the Bristol District of Roads from 1816 to 1824*. Bristol, 1825 (Bristol Record Office; Bristol Public Library).

3. McAdam J. L. *Remarks on the present system of roadmaking*. Longmans, London, 1827.

4. Bentley J. B. and Murless B. J. *Somerset roads: the legacy of the turnpikes. Phase 1: Western Somerset*. Somerset Industrial Archaeology Society, 1985.

5. Bentley J. B. and Murless B. J. *Somerset roads: the legacy of the turnpikes. Phase 2: Eastern Somerset*. Somerset Industrial Archaeology Society, 1987.

18. New Bridge, Bath

HEW 1107
ST 717 658

Until the eighteenth century, the five-span medieval 'Old Bridge' was the only fixed crossing of the River Avon in Bath. Construction of a second bridge, 3 miles downstream on the western outskirts, was started in 1735 and the 'New Bridge' was completed by 1740. Designed by John Strahan and built by Ralph Allen, it is an 85 ft span masonry arch with a circular opening through each spandrel.[1]

It was widened in the early 1830s as part of the turnpike improvements and at the same time the approaches were lengthened and flattened to ease the gradient over the bridge. The new works incorporated a series of arches in the approaches at each end, largely for architectural effect. These give the south elevation, which is that seen from the approach road, an almost monumental effect. Less attention was given to the north elevation.

In the 1960s the sharp bend on the eastern approach was eased by widening on the north side.

The bridge carries the A4 road between Bath and Bristol.

1. BUCHANAN R. A. The bridges of Bath. *Bath History*, **3**, Alan Sutton, Gloucester, 1990.

19. Victoria Suspension Bridge, Bath

HEW 811
ST 741 650

Apart from Clifton Suspension Bridge, the Victoria Bridge is the only survivor of eight suspension bridges of various designs, built across the River Avon in the Bath and Bristol areas in the early decades of the nineteenth century. These designs, very much in the experimental stage, included the conventional chain of links with vertical suspenders (as for Clifton); stayed cantilevers, in which a series of straight ties splayed out from the towers to support the deck at intervals; and James Dredge's patent 'Taper Principle' arrangement typified by the Victoria Bridge.

Dredge formed each link from a number of parallel wrought iron bars. He arranged the chains and inclined wrought iron suspension bars such that the tension in the chains decreased from the towers to mid-span. Consequently the number of bars in the links could be reduced towards mid-span, resulting in a 'tapered' chain.

Victoria Bridge, which was completed in December 1836, has a clear span of 139 ft 10 in. with a deck width of

R. CRAGG

Victoria Bridge, Bath

17 ft 9 in. It was refurbished in the 1940s and again in 1995 and is now used by pedestrians and cyclists.

There are examples of Dredge's design, differing in detail, at Stowell Park, Wiltshire;[1] Aberchalder, Inverness-shire;[2] and at Caledon and Castledawson in Northern Ireland.[3]

1. Stowell Park Suspension Bridge (HEW 378) SU 146 614.

2. Victoria Suspension Bridge, Aberchalder (HEW 888) NH 337 036.

3. McQuillan D. From brewer to bridge builder: reflections on the life and work of James Dredge. *Proc. Instn Civ. Engrs, Civ. Engng*, 1994, **102**, Feb., 34-42.

20. Bath Green Park Station

HEW 665
ST 745 648

This station has been disused for railway purposes since 1966 but the building has now been restored to its original condition to form part of a supermarket which a national grocery firm has developed on the site.

Opened on 1 May 1870 as the terminus of the Midland Railway branch from Mangotsfield on the Bristol to Gloucester line, the station was used jointly from 1874 as their Bath terminus by the Somerset and Dorset Joint Railway.

Green Park station is outstanding among the lesser Victorian stations. The Georgian frontage, with Ionic columns above the rusticated ground floor, a balustraded parapet, well-proportioned fenestration and a delicate iron *porte-cochère* in front of the entrance, is well in keeping with the architecture of Bath.

The train shed behind has 14 bays of simple wrought iron girder arches of 66 ft 6 in. span, flanked by narrow aisles of segmental cast-iron arches, all supported on slightly tapering octagonal cast-iron columns. It is a good example of the pleasing effect of straightforward engineering built without extraneous decoration but to the right proportions.

An unusual feature of the station was the series of basements under the platforms and station buildings. These were used as bonded stores and connected by underground passages to a Customs building which stood at the west end of the station yard.

21. Bath Roman Water Supply and Drainage

HEW 1019
ST 752 648

Few visitors to the Roman Baths at Bath realize what lies beyond the steamy cavity from which emerges the hot water supplying the Great Bath.[1]

The original developers, in the first century AD, impounded the hot springs in a roughly octagonal reservoir about 1600 sq. ft in extent, made by raising ashlar masonry walls to a height of about 8 ft. To get rid of water during construction, openings were left, later to be plugged by 1 ft square oak blocks, some of which were visible when the reservoir was discovered in 1878.[2,3] The floor and walls were lead-lined. A lead pipe 20 in. wide and 5 in. deep fed the Great Bath, which in turn fed two smaller baths.

Surplus water from the reservoir and the overflow from the baths, totalling about 300 000 gallons each day, was drained to the River Avon 400 yd away in a 3 ft span arched masonry culvert with a timber-lined channel at the bottom. For ease of access to repair and clean this drain, it was furnished with manholes at suitable intervals and was built high enough for a man to walk through upright. Much of it remains in use.

In the second and third centuries extensive alterations

and extensions were carried out, involving more drainage and pipework. Some of the pipework can still be seen and much of the remainder can be traced by the recesses in which it lay in the paving round the baths.

1. Cunliffe B. *Roman Bath discovered.* Routledge and Kegan Paul, London, 1984, 109–148.

2. Mann R. Note on a Roman culvert. *Proc. Br. Archaeolog. Assoc.*, 1878, **34**, 15 May, 246–248.

3. Irvine J. T. Discovery of a Roman reservoir at Bath. *Proc. Br. Archaeolog. Assoc.*, 1882, **38**, 4 Jan., 91–93.

22. Pulteney Weir, Bath

HEW 1272
ST 752 650

In medieval times a ferry crossed the Avon in Bath at the site now occupied by Pulteney Bridge, the only bridge in Britain carrying buildings on both sides. Just below this, a weir ran diagonally upstream from the right (west) bank and turned sharply in horseshoe form towards the left bank. At each end of the weir was a water-mill.

The weir remained in much the same form after the mills had disappeared and after the building of the bridge in 1774.

Over the centuries Bath was subject to floods, aggra-

Pulteney Weir, Bath

R. CRAGG

vated by the presence down river of many weirs. In 1727 the Avon was made navigable as far up as Bath by means of six locks, for 74 ft by 16 ft barges, in the 11 miles from Hanham (ST 647 700), but this did nothing to improve the flooding situation.

In 1824 Thomas Telford was consulted and proposed channel improvements but surprisingly made no mention of the weirs. Nothing was done and this was the fate of several quite ambitious proposals over the next 130 years.

In 1965 the Bristol Avon River Authority undertook over £2 million of improvement works through Bath from Pulteney Bridge to Saltford (ST 690 669) and about a tenth of the cost was for a new weir at Pulteney Bridge.[1] This is a sharply pointed horseshoe of three broad concrete steps springing from the right bank of the river and terminating downstream in a boat-shaped artificial island; between it and the left bank is the flood discharge channel controlled by a radial sluice gate.

The site of this imaginative scheme, in the midst of the architectural gems of Bath, is aesthetically sensitive. With the advice of Sir Hugh Casson the designers won a Civic Trust Award. Judicious landscaping has blended the sluice structure, very much in the modern idiom, into the environment, while the Award described the weir as a visual and acoustic triumph, with three great stepped crescents of foam contrasting with the quiet waters above them reflecting the Pulteney Bridge.[2]

1. GREENHALGH F. *Bath flood prevention scheme.* Wessex Water Authority, Bath, 1974.

2. Pulteney Bridge, Bath (HEW 1060) ST 752 650.

23. Kennet and Avon Canal

HEW 1034
ST 754 644 to SU 470 672

The Kennet and Avon Canal, built between 1794 and 1816 with John Rennie as Engineer, runs for 57 miles from the River Avon at Bath to the Kennet Navigation at Newbury, via Devizes and Hungerford. It afforded a more usable route between London and the Severn than the Stroudwater and the Thames and Severn Canals.

The opening of the Great Western Railway in 1841 rapidly affected its fortunes and by the early part of the twentieth century it had largely fallen into disuse. However, through navigation from Bristol to London has been made possible again by an ambitious restoration programme, culminating in the 1990 reopening of the Caen Hill flight of locks at Devizes.

The canal was constructed 44 ft wide at the surface with locks about 75 ft by 14 ft. It leaves the Avon by way of Widcombe Locks, Bath. Nearby was the site of Dolemeads Wharf to which stone was brought down from Combe Down quarries, 1½ miles to the south-east of Bath, by Ralph Allen's Waggonway.[1] Built in 1730, this waggonway was probably the first railroad in Britain to have been described in print, having been the subject of an article by Charles de Labelye in Desagulier's *Experimental Philosophy.*[2]

After the top lock of the Widcombe flight, the canal passes through the ornamental Sydney Gardens, with a very short tunnel at each end and two decorative cast-iron bridges crossing over, as it follows the Avon valley towards Melksham. Claverton Pumping Station supplies water to this 9 mile long pound and the canal twice crosses the Avon, first by means of Dundas Aqueduct and secondly by the Avoncliff Aqueduct (ST 804 600).

Thirty-seven locks in a length of 12 miles then raise the canal a total of 287 ft to a 15 mile long level pound along the Vale of Pewsey. The canal is lifted by a further four locks at Wootton Rivers to the 2½ mile long summit level, which is 404 ft above the Avon in Bath. This level, which is supplied with water by the Crofton Pumping Station, passes through the 502 yd long, brick-lined Bruce Tunnel.[3] The canal then falls through a total of 210 ft to Newbury by means of a further 31 locks.

There are numerous bridges in timber, cast iron, wrought iron, brick and masonry, notable among the last being the decorated Ladies Bridge[4] at Wilcot. Some of the swing bridges still have ball bearing pivots—as at Bathampton, Allington and Hungerford.

At Great Bedwyn, the Kennet and Avon Canal is crossed by an arch[5] of about 25 ft 3 in. span carrying an unclassified road. Said to be one of the first skew arches built in brick, it has certain irregularities in the coursing of the brickwork which have led to the charge that John

R. CRAGG

Great Bedwyn Skew Bridge, Kennet and Avon Canal

Rennie, the designer, had not fully grasped the principles. However, like the non-skew bridges along the canal, the width is greater at the springing than at the crown and this feature would have introduced problems in the layout of the bricks as compared with the case of a parallel ended barrel.

The suspension bridge in Stowell Park[6] is by James Dredge, to the same patent design as his Victoria Bridge, Bath.

1. Ralph Allen's Waggonway, Bath (HEW 1338) ST 752 643 to ST 758 622.

2. DESAGULIER J. T. *A course of experimental philosophy*. Printed for J. Senex *et al.*, London, 1734, **1**, 274–279 with 3 plates.

3. Bruce Tunnel (HEW 1077) SU 236 632.

4. Ladies Bridge, Wilcot (HEW 895) SU 135 610.

5. Great Bedwyn Skew Bridge (HEW 379) SU 276 637.

6. Stowell Park Suspension Bridge (HEW 378) SU 146 614.

24. Widcombe Locks, Bath

HEW 1110
ST 754 643 to ST 758 646

Widcombe Locks lift the Kennet and Avon Canal through 66 ft 6 in. from its junction with the River Avon at Bath to the side slopes above the river. Originally there were seven locks, but now there are six, each about 75 ft long and just over 14 ft wide, built to take 50 ton barges, 70 ft long and 13 ft wide. Each lock is furnished with a side pond dug out of the hillside.

They were begun in Bath stone against Rennie's advice, but were completed in a more durable stone from Winsley quarry. The locks have been well maintained with repairs done in red or blue engineering bricks. In 1973 the two bottom locks were converted to a single deep lock during major road alterations.

The hand-operated timber gates have built-in sluices, those in the lowest gates having been made hydraulically assisted during the lock conversion.

A charming feature of the locks is the pair of tiny cast-iron footbridges, over the Top Lock[1] and Wash House Lock,[2] built by Stothert of Bath circa 1815.

A similar bridge[3] of 27 ft span carries a footpath over Kings Weston Lane in Bristol.

1. Top Lock Bridge (HEW 360) ST 758 646.

2. Wash House Lock Bridge (HEW 806) ST 757 644.

3. Kings Weston Lane Bridge (HEW 649) ST 545 773.

25. Cast-iron Bridges, Sydney Gardens, Bath

A short way northward along the Kennet and Avon Canal from Widcombe Locks may be seen three more cast-iron arched bridges, the first two over the canal and the third over the adjacent railway.

HEW 808
ST 759 654

The first is a footbridge of about 23 ft span and 10 ft wide, having four cast-iron ribs in two sections with cast-iron deck plates. The ribs have ring pattern spandrel filling. The fencing is of simple palings with three rails, alternate palings being full height and three-quarter height. This bridge was overhauled in 1978.

HEW 807
ST 759 653

The second is very different. It spans just over 30 ft on a pronounced skew and carries a 19 ft wide roadway. The

R. CRAGG

Cast-iron canal bridges, Sydney Gardens, Bath

face ribs are solid panelled in two sections, and there are five interior ribs of T or cruciform section carrying solid cast-iron deck plates 10 ft wide set parallel to the canal. The parapets have distinctive cast-iron panels of a large diamond superimposed on a cross.

The ironwork for both these bridges, which date from about 1800, was cast at Coabrookdale. Access to them is from Sydney Road alongside a building erected above Sydney Gardens No. 1 Tunnel (and once housing canal offices) and back under the building through the tunnel.

HEW 359
ST 758 653

The third bridge is much later, built in about 1865 over the Great Western Railway. The span of 30 ft 8 in. reflects the width of the railway formation required to accommodate Brunel's broad gauge tracks. Some 50 yd away towards Bath a skew masonry bridge carrying a road over the railway bears a plaque to 'Isambard Kingdom Brunel 1841'.

26. Claverton Pumping Station

HEW 1076
ST 791 644

Claverton water-powered pumping station started work in March 1813 lifting water from the River Avon to the Kennet and Avon Canal 53 ft above.

Originally, a breast wheel, 25 ft wide and 18 ft 4 in. diameter, which was fed through a leat from the river, operated the two bucket pumps through two gear wheels to a common crankshaft from which in turn cruciform section connecting rods work the pumps. The larger gear wheel has wooden teeth. In 1858 the Great Western Railway, who then owned the canal, replaced the single wheel by a pair, each 15 ft 6 in. diameter and 11 ft wide but still on one shaft, and introduced an intermediate bearing on a cast-iron A-frame.

In 1952 some of the wooden teeth stripped and a small diesel pump was installed to maintain the water supply.

The pumps are housed in a masonry building, with the water-wheels under a timber structure with a tiled roof.

Restoration of the machinery has been carried out by the voluntary efforts of Bath University and the Kennet and Avon Canal Trust. In 1981 the British Waterways Board installed two 75 h.p. electric pumps just upstream of the station and presented the diesel pump to the Trust, for preservation.

Claverton Pumping Station

R. CRAGG

R. CRAGG

Dundas Aqueduct

HEW 188
ST 785 625

27. Dundas Aqueduct

The Dundas Aqueduct lies in the Limpley Stoke Valley, south-east of Bath, and carries the Kennet and Avon Canal over the River Avon. There are right angled bends in the canal at each end and a turning basin, which also served as the terminal basin of the long defunct Somerset Coal Canal, was provided at the Bath or west end. The main semicircular 65 ft span stone arch over the river is flanked by two tall semi-elliptical arches of 19 ft 3 in. span.

The aqueduct was begun in 1796, completed in 1798 and opened in 1800. No one would tender for the foundations, which were built in coffer-dams by direct labour. James McIlquham was the contractor for the superstructure.

With its exterior cornices on both sides and stone balustrade, the aqueduct forms a monumental piece of architecture. It is a fine example of the work of John Rennie, who surveyed the route of the canal from Bath to Newbury in 1788, at the age of 29, and was appointed Engineer for the canal in 1794.

The aqueduct is named after Charles Dundas, Chairman of the Canal Company from its inception until his death, as Lord Amesbury, in 1832.

Serious leakage through the aqueduct was remedied in the 1980s by installing a concrete lining. Similar measures were necessary at the Avoncliff Aqueduct.

Nearby, the former entrance lock (now without gates) of the Somerset Coal Canal and the first quarter mile of the canal were restored in 1986–88 and are now used as private boat moorings.

28. Devizes Locks

HEW 1078
ST 966 617 to SU 000 617

West of Devizes, 29 locks raise the Kennet and Avon Canal from Lower Foxhangers Lock 22, west of the B3101 near Rowde, to Devizes Top Lock 50 at Town Bridge. They are to be compared with the locks on the Worcester and Birmingham Canal at Tardebigge but although this has the greatest number of locks in one flight in Britain,

Devizes Locks, Caen Hill Flight

R. CRAGG

the Devizes locks rise through a greater height (237 ft) over a lesser distance (about 2¼ miles).

Locks 28 to 44 form the remarkable Caen Hill flight, where 17 locks, on a gradient of 1 in 30, are separated from one another only by the entrances to the closely spaced side ponds on the north side of the canal.

In 1829, some years after the opening of the locks, gas lighting was installed and an extra charge was made for craft passing through at night with the aid of the lighting.

The locks were closed in 1951, but were reopened by Her Majesty Queen Elizabeth II on 8 August 1990, following major restoration works carried out as a joint endeavour between Kennet District Council, the British Waterways Board and the Kennet and Avon Canal Trust. A pumping station has also been built to pump water from the bottom lock back up to the top of the Caen Hill flight.

On the west side of Prison Bridge, Devizes, is a tablet commemorating John Blackwell, one of Rennie's resident engineers, who went on to serve the canal company as Superintending Engineer for 34 years until his death in 1840.

29. Crofton Pumping Station

HEW 57
SU 262 622

Situated about 2 miles west of Great Bedwyn, this pumping station was built to feed water to the summit level of the Kennet and Avon Canal. Water is pumped from Wilton Water, some 300 yd to the south, three-quarters of a mile to a point just west of Crofton top lock. In contrast with Claverton Pumping Station, it was steam-powered.

The building houses two nineteenth-century engines. The 1812 engine, 42 in. bore and 8 ft stroke, is the only Boulton and Watt beam engine still in steam and performing the work for which it was installed. The Smethwick engine, housed in the Museum of Science and Industry in Birmingham, is older (1779) and is still steamed but no longer on its original site. The second engine, a Sims combined cylinder engine, 40 in. and 22 in. bore and 8 ft stroke, built by Harveys of Hayle, was installed in 1846 to replace an even earlier Boulton and

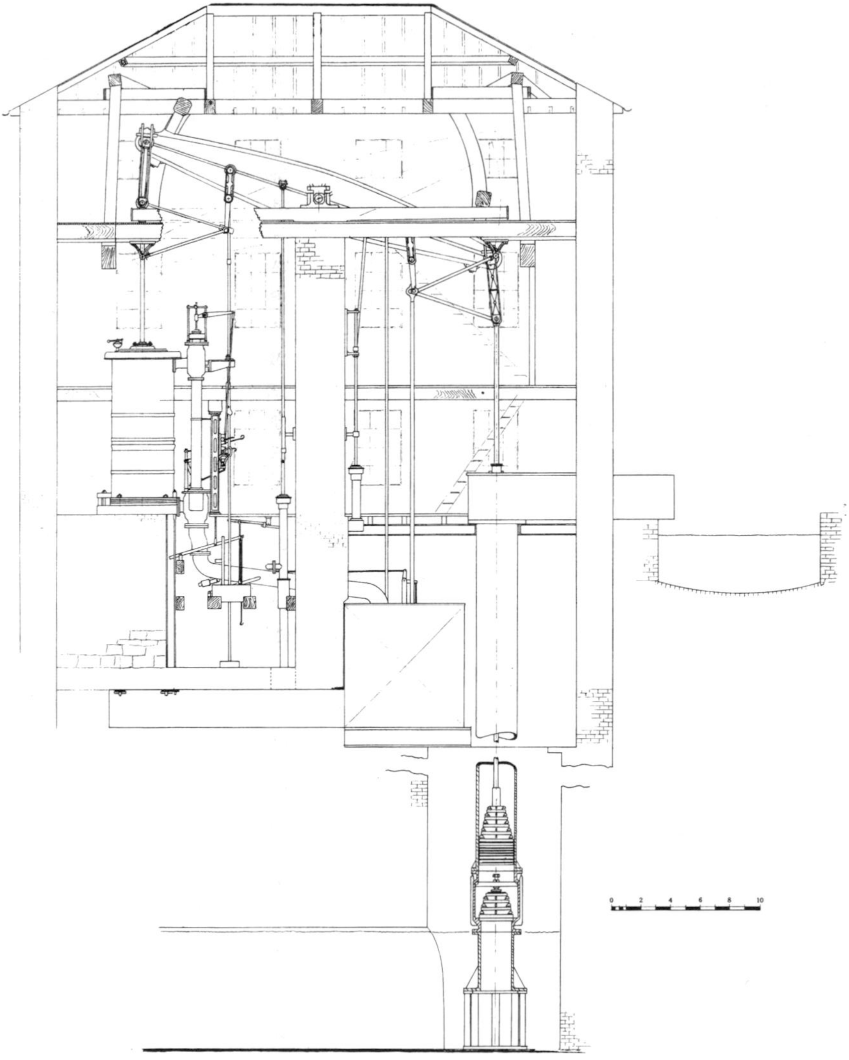

Watt engine of 1809. The Sims engine was rebuilt as a single cylinder engine at the turn of the century by the Great Western Railway.

Both engines have been restored by the Kennet and Avon Canal Trust and operate on several occasions each year when the station is open to the public.

Crofton Pumping Station: Boulton and Watt 42 in. engine

30. Silbury Hill

HEW 510
SU 100 685

Adjoining the north side of the A4, some 6 miles west of Marlborough, Wiltshire, and close to the Avebury Stone Circle, is the largest prehistoric man-made mound in Europe. Dating from 2660 BC, it is roughly 430 ft in diameter and 130 ft high with side slopes of 1 in 1.5, and it is surrounded by a 30 ft wide ditch. Exploratory tunnelling over the past 200 years has established that it is not a burial mound as was once thought. Although its purpose remains unknown, it has been shown that the construction was quite sophisticated for its time.[1]

Over and around an original 100 ft diameter mound was built a series of radial and concentric walls of chalk blocks which were filled in with chalk rubble to form a number of 17 ft high steps and benchings. These were then packed with rubble and the whole mound turfed over.

The borrow ditch left the base with a 30 ft high vertical face of virgin chalk, against which was piled chalk rubble in 2 ft deep stepped layers with timber revetment, to protect the face against weathering and slipping. This protection is now buried in the silt which has partly filled the ditch over the centuries.

Silbury Hill

B. T. BATSFORD

Even by modern standards it was an impressive undertaking. It has been estimated that 500 people, with the most primitive digging tools and carrying material in baskets, would have taken ten years to complete the work and that in relation to the likely available population the enterprise was, for its time, comparable to the American or Russian space programmes.

1. DANES M. *The Silbury treasure*. Thames and Hudson, 1976, chapters 1–3.

31. Blackland Mill Iron Roof

HEW 1702
SU 017 693

In 1812 a Gloucestershire iron manufacturer, advertising his recently patented wrought iron system for roofs, floors and ceilings, emphasized its advantages over conventional timber systems in terms of durability, lightness, cheapness and 'perfect security from fire'. The patentee, Thomas Pearsall of Willsbridge, Bitton, gave as an example the roof of Blackland Mill, near Calne, completed in 1811.

This is one of the earliest surviving wrought iron roofs in Britain. With a span of 17 ft and 58 ft 4 in. long, it is constructed from 2⅞ in. deep by ⅛ in. thick 'hoop-iron' rafters at 10 in. centres, pitched at 45° and bolted at the apex. Battens 1 9/16 in. by 1/12 in. are halved into the rafters at 9 in. centres and secured with iron clips. The rafters are bedded on limestone wall plates and the outward thrust is counteracted by 1⅝ in. by 1/16 in. vertical hangers and inclined ties at 20 in. centres.

Currently, this building is not open to the public.

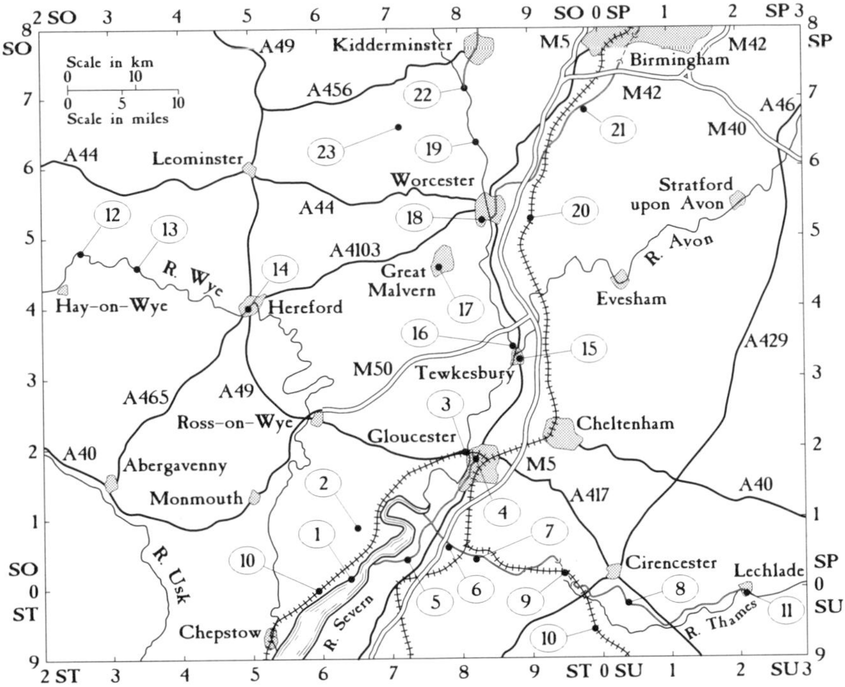

1 Lydney Docks
2 Roman Road, Blackpool Bridge, Forest of Dean
3 Over Bridge, Gloucester
4 Gloucester Docks
5 Gloucester and Sharpness Canal
6 Stroudwater Canal
7 King's Stanley Mill, Stonehouse
8 Thames and Severn Canal
9 Sapperton Tunnel, Thames and Severn Canal
10 Railway from Swindon to Severn Tunnel Junction via Gloucester
11 Town Bridge, Lechlade
12 Whitney Toll Bridge
13 Bredwardine Bridge
14 Wye Bridge, Hereford
15 Quay Pit Bridge, Tewkesbury
16 Mythe Bridge, Tewkesbury
17 Great Malvern Station
18 Powick Bridges, Worcester
19 Holt Fleet Bridge
20 Birmingham and Gloucester Railway
21 Tardebigge Locks
22 Stourport-on-Severn Canal Terminus
23 Stanford Bridge

5. Gloucestershire, Hereford and Worcester

This region was not greatly affected by the Industrial Revolution which from the 1760s developed rapidly in the West Midlands just to the north. Consequently main towns such as Hereford, Evesham and even Gloucester still perform their ancient functions as commercial centres for a predominantly rural area. However, timber, iron ore and coal, mined until recent years in the Forest of Dean, supported some local metal working industry. They were also exported to other parts of the country through the small port of Lydney.

The first brass making in Britain was set up in the Forest in the sixteenth century before being transferred to Bristol. South of Gloucester a number of textile mills were built to service the once extensive woollen industry based on Cotswold sheep. Some of these have survived, now converted for other uses.

Communications from earliest times were dominated by the River Severn running from north to south through the area. Development of the industrial Midlands required links from the river to the higher ground to the north-east. The Staffordshire and Worcestershire Canal (1772) and the Worcester and Birmingham Canal (1815) are typical early examples of these links. The Stroudwater and the Thames and Severn Canals aimed at joining the Severn and the Thames but were largely superseded by the Kennet and Avon Canal further south. The Gloucester and Sharpness Canal, by cutting off a dangerous loop of the Severn, brought about rapid expansion of Gloucester as a port, followed late in the nineteenth century by Sharpness Docks.

After the canals came the railways, with trunk routes running north and south and with east to west lines leading into Wales. In more recent times the M5 and M50 motorways have introduced a new element into the landscape.

It is not surprising therefore that many of the interesting civil engineering works in this area are related to transport by one means or another.

1. Lydney Docks

HEW 626
SO 634 018 to
SO 651 104

Lydney was a port from the twelfth century but over the centuries siltation of its small river gradually reduced its usefulness.

In 1810 the Severn and Wye Tramroad Company, which operated 30 miles of tramroad in the Forest of Dean, built an entrance lock from the Severn and a basin 760 ft long and 105 ft wide, with a canal 3050 ft long leading to an upper basin 1100 ft long and 90 ft wide to take 100 ft by 24 ft barges. In 1821 a tidal basin and outer lock were added, which enabled ships up to 400 tons to use the lower basin. So Lydney became the principal sea outlet for coal mined in the Forest, and at their peak the docks handled as much as 400 000 tons in a year.

Ownership passed to the Great Western Railway and the Midland Railway jointly in 1894 and, after nationalization of the railways in 1948, to the British Transport Commission and eventually to the British Transport Docks Board. With the cessation of mining in the Forest, the coal trade disappeared and, of the banks of coal sidings which ran along the docks to serve nine coaling stages, there is now little trace.

Severn Trent Water Authority purchased the docks in 1980 to prevent their irreversible closure. Subsequently the structures, buildings and docks at the lower end were designated a scheduled ancient monument.

In 1989 ownership was transferred to the National Rivers Authority, now part of the Environment Agency, which has set about a programme to develop the amenity potential of the docks.

2. Roman Road, Blackpool Bridge, Forest of Dean

HEW 630
SO 653 087

This fragment of the Roman road between Lydney and Mitcheldean (Ariconium) was built in the first or second centuries AD. It lies alongside an unclassified road off the Blakeney to Parkend road, about 2 miles from Blakeney and leading northward to Upper Soudley.

The remains extend for about 85 ft and include part of a road junction. Between two lines of edging stones 4 in.

to 6 in. wide and 12 in. to 18 in. long is random paving of rough stones up to about 18 in. by 12 in. in plan. The width, just under 8 ft, suggests that the road was an *actus*, or a single carriageway. This was the second class of Roman road, the first class being the *via*, a two-lane road 14 ft or more wide.

The Romans built their roads with military considerations in mind, in straight lines from one vantage point to another, a strip on each side being cleared of trees and bushes to guard against marauders. Construction was in layers, usually on an embankment, the *agger*, dug from borrow ditches or pits alongside. The material used depended on what was locally available. The thickness could vary from 3 in. up to 2 ft according to the weight of traffic expected. On the most heavily used roads there would be a layer of small stones or gravel, followed possibly by stone-lime concrete or, in ironworking districts, iron slag, worked into a firm mass. Finally the wearing surface, laid between edging stones, consisted of roughly rectangular or polygonal stones, several inches thick, closely fitted together. The road surface was therefore raised above the surrounding ground and with the heavy camber provided, as much as a foot in a width of 14 ft, formed a well drained highway.

It is interesting to compare the paved surface of this Roman road with that of a short length of paving, said to be medieval, which is preserved opposite Minsterworth Church, about 9 miles north-east of Blackpool Bridge.

3. Over Bridge, Gloucester

HEW 148
SO 817 196

Immediately to the south of the present bridge which carries the A40 trunk road across the Severn about 1 mile west of Gloucester, a masonry arch of 150 ft span crosses the river.[1] It was completed in 1828. Thomas Telford was the Engineer,[2] and he based the detailed design on Perronet's five-span bridge across the Seine at Neuilly (1774, replaced in 1956).

The main soffit is built to an elliptical curve with 35 ft rise, but the voussoirs on the face follow a segmental curve of only 13 ft rise and spring 22 ft above the main springing. This produces a funnelled entry upstream and

R. CRAGG

Over Bridge

downstream aimed at easing the passage of flood water through the bridge, and at the same time gives an interesting complex shape to the arch.[3]

The abutments rise from timber platforms lying on gravel 27 to 33 ft below ground level. The wing walls are founded 8 to 10 ft deep, directly onto the ground.

When the arch centring was struck, the crown of the arch sank 2 in., and later movement of the wing walls produced a further sinking of 8 in. Telford blamed this on his parsimony in omitting piling or platforms under the walls and considered the work as being one of his failures.

Nevertheless, the bridge, which had a 17 ft roadway and two 4 ft footways, carried traffic on the A40 until it was replaced by the new bridge upstream in 1975.

It now stands in isolation, as a scheduled ancient monument.

1. HEYMAN J. and THRELFALL B. D. Two masonry bridges: II—Telford's bridge at Over. *Proc. Instn Civ. Engrs*, 1972, **52**, 319–330.

2. RICKMAN J. (ed.) *Life of Thomas Telford ... with a folio atlas of copper plates.* London, 1838, plate 66.

3. The new bridge at Over. *Glos. J.*, 1828, 24 May, 3.

4. Gloucester Docks

HEW 625
SO 827 183

Gloucester has been a river port since Roman times. Its twelfth- and thirteenth-century developments are remembered by the street known as The Quay.[1,2]

South of this the present docks comprise the Main Basin (1812), a Barge Arm (1824), two dry docks (1834 and 1851) and the Victoria Dock (1849) on the east side of the basin, and Monk Meadow Dock (1890) on the west side of the Gloucester and Sharpness Canal. The canal, which enters the south end of the basin, is widened over its last 1½ miles to afford berthing facilities. At first, access to the docks was solely through a lock from the Severn at the north-west corner of the Main Basin, but with the opening of the canal in 1827, the use of the lock became confined to barge traffic to and from the upper reaches of the river.

The docks are notable for nearly a score of multi-storey warehouses, the majority built in the first half of the nineteenth century to meet the rapid increase in the trade in imported corn which followed the relaxation and subsequent repeal of the Corn Laws.[3] The earliest dates from 1827. All follow the same general pattern: brick walls and

Gloucester Docks

RUSSELL ADAMS FRPS

slate roofs, with timber floors carried on rows of cast-iron tubular columns, loading doors on all floors and hand-operated hoists projecting from the roofs. The Pillar Warehouse (1849), a Grade II listed building on the east side of the canal just south of the docks, is of particular interest, with the three upper floors on the waterside hanging over the quay and supported on a colonnade of seven cast-iron Doric columns.

Although traffic declined steadily during this century, the docks continue to trade and the owners, British Waterways, are cooperating with the City Council to preserve the docks and refurbish the warehouses, not only as a commercial feature of the city, but also as a centre for leisure, cultural and business activities.

1. Brewster R. S. *The port of Gloucester its history and development from the earliest settlement to 1976.* (Typescript 1978 in Gloucester Public Library.)

2. Conway-Jones A. H. *Gloucester Docks, an illustrated history.* Alan Sutton and Gloucester County Library, 1984.

3. Conway-Jones A. H. The warehouses at Gloucester Docks. *J. Glos. Soc. Ind. Archaeol.*, 1977–78, 13–19.

5. Gloucester and Sharpness Canal

HEW 466
SO 827 185 to
SO 668 031

The 16½ mile long Gloucester and Sharpness Canal (formerly the Gloucester and Berkeley Canal) was the second ship canal to be built in Britain, the first being the seventeenth-century Exeter Canal.[1] It was built between 1798 and 1827 to the design of Robert Mylne, with Thomas Telford[2] as adviser to the Canal Commissioners. The Pinkerton family, who worked on canals and navigations from Yorkshire to South Wales,were involved as contractors as was also Hugh McIntosh. Eighteen feet deep and 86 ft 6 in. wide, it was meant for 400 ton ships, but modern ship design permits the passage of ships up to 1200 tons, with a maximum beam of 29 ft which is determined by the bridge openings.

The canal connects with the Severn through a lock at Gloucester Docks. At Sharpness a sea lock leads to a basin from which two locks, the smaller of which is now sealed, lead to the canal. The basin and lock structures are basically as originally built and there is an attractive Lock-master's House.

In 1874 a new dock at Sharpness[3] was opened. Entrance gates 57 ft wide lead to the Tidal Basin from which a 320 ft by 57 ft lock, with 22 ft of water on the cill, connects to the 20 acre dock, some 2000 ft long, which can accommodate 7000 ton ships. At the inner end of the dock a junction was made with the canal, and canal traffic now uses the dock entrance. The canal from this junction to the original basin is used as a marina.

Less than a mile north-east of the dock, at SO 679 033, stand the masonry pivot tower and east abutment of the swing bridge built in 1874–76 to carry a railway across the canal as part of the Severn Railway Bridge.[4] This was one of the longest railway bridges in the United Kingdom and carried a single track to link the Birmingham to Gloucester and Bristol line of the Midland Railway to the Great Western line from Gloucester to Chepstow and South Wales.

There were 21 spans of wrought iron bowstring trusses, ranging from about 125 ft to 325 ft long, on cast-iron cylindrical piers, together with a twelve-arch masonry viaduct at the west end and the swing bridge over the canal at the east end. Two spans of the bridge were brought down in 1960 by a petrol barge which collided with one of the piers in fog and eventually the whole structure was demolished in 1968–69.

At Saul Junction (SO 756 093) the crossing of the Stroudwater Canal is of interest, and is substantially as it was in 1820. There is a small dry dock, built in 1827, which is still used by a firm engaged on the repair and building of small craft.

All 15 swing bridges and the lock gates are of recent date but several of the original elegant Adam-style bridge keepers' cottages still remain. Two of the three swing bridges which have been mechanized are now remotely controlled.

1. OTTER R. A. *Civil engineering heritage: Southern England*. Thomas Telford, London, 1994, 74–76 (HEW 529).

2. CRAWFORD G. N. Thomas Telford and the Gloucester and Berkeley Canal. *Industrial Archaeology Review*, 1989, **XI**, 2, 155–170.

3. Sharpness Docks (HEW 388) SO 673 024.

4. Severn Railway Bridge (HEW 338), formerly at SO 679 033 to SO 670 041.

6. Stroudwater Canal

HEW 292
SO 751 104 to
SO 848 050

This canal is of interest because the company formed in 1730 to promote it can lay claim to being the oldest canal company still in being, as although the canal was abandoned in 1954, a small revenue continues to be drawn from fishing rights and water sales.

Opposition from mill owners on the River Stroudwater (or Frome) delayed the start until 1775. However, between 1759 and 1763 there was an attempt, which proved too costly in operation, to form a navigation of the lower reaches of the river and to bypass the mills by the novel device of making the new channel with dead ends overlapping the millponds, and transferring goods from one section to the next by crane.

The 8¼ mile canal, completed in 1779 to plans by Thomas Yeoman, ran parallel to the river from Framilode on the Severn to Wallsbridge in Stroud whence it was later continued by means of the Thames and Severn Canal. It was 42 ft wide and 6 ft deep, and twelve locks dealt with the total rise of 102 ft.[1]

A good deal of the waterway remains, but few structures, of which the most interesting is the lock at Saul Junction, built in 1820 to take the canal across the Gloucester and Sharpness Canal.

The Cotswold Canals Trust (formerly the Stroudwater and Thames and Severn Canal Trust) is engaged in a long-term programme aimed at restoring this canal and the Thames and Severn Canal to navigation between Saul Junction and the Thames.

1. HANDFORD M. *The Stroudwater Canal.* Alan Sutton, Gloucester, 1979.

7. King's Stanley Mill, Stonehouse

HEW 426
SO 813 043

Near the Stroudwater Canal, this five-storey textile mill,[1] built in 1813, produced fine woollen cloth and industrial fabrics until 1989. There is a continuous record of mills on the site going back to 1563, with a mention in the Domesday Book.

It differs from other mills in the area in that its walls above the first floor are of brickwork, instead of being in Cotswold stone throughout. The elevations include a number of handsome Palladian-style cast-iron venetian windows.

The building is Z-shaped in plan. The eastern section was built with conventional timber floors. However, the western and central sections were built to be completely fireproof and incorporated no combustible material; they even had iron doors.

Internally the floors of the fireproof sections are carried on arcades of cast-iron columns which support an elaborate system of ornate cast-iron arches. These in turn carry cast-iron floor beams between which are brick jack arches. Over these is a filling of ash or rubble on which the stone floor slabs are laid.

The ironwork was cast at Dudley in the West Midlands and brought to the site along the Stroudwater Canal whence it was hauled for a distance of about ⅓ mile over the fields. In view of their size and number, the accuracy of the castings is remarkable, in that the seatings for the bearings of the complex system of line shafting are integral with the castings, with limited scope for adjusting the alignment.

The original five breast-shot water-wheels were supplemented with a 40 h.p. steam engine in 1840. Two of the wheels were replaced in 1867–68 with an 80 h.p. water turbine, later converted to generate electricity. The wrought iron roof structure was fire damaged in 1884 and reconstructed in timber.

It is intended that a refurbishment programme dealing with the upper parts of the walls and the roof, together with adjacent historically important buildings, will be followed by restoration to working order of a selection of the original machinery.

1. TANN J. *Gloucestershire woollen mills*. David and Charles, Newton Abbot, 1967, 149–151.

8. Thames and Severn Canal

HEW 1139
SO 848 050 to SU 205 988

Completed in 1789 to the design of Robert Whitworth with Josiah Clowes as Engineer in charge, this canal joined the east end of the Stroudwater Canal to the Thames just above Lechlade, and completed the first waterway link between Thames and Severn.[1] There were

44 locks in its 28¾ mile length. At its west end, Severn trows could reach Brimscombe where there were quays and warehouses and facilities for weighing boats. The rest was for narrow boats only.

At Thames Head (SU 988 988) the canal was fed from the springs at the source of the Thames, pumped up first by wind power, then by a Watt steam engine installed in 1802 and replaced in 1854 by a Cornish engine.

The canal was abandoned in 1927–33 and although much has disappeared some remains can be seen, including a number of locks and overbridges and the portals of Sapperton Tunnel. The Cotswold Canals Trust has embarked on a long-term programme aimed at restoring to navigation the whole line of the canal.

There are interesting canal keepers' cottages, still lived in, at Cerney Wick (SU 079 960), Marston Meysey (SU 132 963) and Chalford and Inglesham Lock on the Thames (SU 205 988). These are circular, three-storey stone buildings: a store and a stable on the ground floor; a 16 ft 10 in. diameter living room reached by outside steps; stairs between the inner and outer walls leading to a bedroom; and a roof in the form of an inverted cone designed to catch rainwater.

1. Household H. *The Thames and Severn Canal.* Alan Sutton, Gloucester, 1983, 2nd edn.

9. Sapperton Tunnel, Thames and Severn Canal

HEW 629
SO 943 033 to SO 966 006

Sapperton Tunnel lies under the A419 Stroud to Cirencester road, between Sapperton and Coates in Gloucestershire.

Built between 1783 and 1789 on the summit level of the canal, it was at the time the longest canal tunnel in the country, being 2 miles 297 yd long.

Both portals, easily accessible, lie in pleasant wooded valleys. The work was designed by Robert Whitworth assisted by Josiah Clowes[1] and was built by five separate gangs of workmen. Difficulties were encountered in the building of the tunnel and it was here that Clowes established his reputation as a canal engineer. The east portal, restored by the Stroudwater and Thames and Severn

R. CRAGG

East portal, Sapperton Tunnel

Canal Trust (now the Cotswold Canals Trust), is a striking structure in Cotswold stone in the classical style. The Trust now proposes to restore the west portal which is of plain limestone masonry.

There was no towpath and boats were propelled by the crew lying on their backs and pushing against the sides and roof of the tunnel with their feet, a technique known as 'legging'.

Access right through the tunnel is now no longer possible because of roof falls. However, boat trips run from the eastern end for some 1000 ft.

1. LEWIS C. Josiah Clowes (1735–1794). *Trans. Newcomen Soc.* 1978–79, **50**, 155–158.

10. Railway from Swindon to Severn Tunnel Junction via Gloucester

HEW 1179
SU 146 851 to
ST 460 875

The area covered by this chapter is traversed by much of the route of this railway line. It began as the Cheltenham and Great Western Union Railway,[1] a broad gauge line authorized in 1836, engineered by I. K. Brunel and built under the direction of his chief assistant, R. P. Brereton.

The line diverges northward from the London to Bristol railway a short way west of Swindon station. At SO 960 013 it crosses the line of the Sapperton canal tunnel on the Thames and Severn Canal and soon after come the Sapperton railway tunnels.[2] These were originally envisaged as one curved 2800 yd long tunnel at low level. However, nine shafts had been sunk and the pilot heading nearly completed when, in 1841, financial stringencies obliged Brunel to take the drastic action of raising the track some 90 ft. This resulted in two tunnels separated by a very short cutting near the A419 Stroud to Cirencester road at SO 946 019.

Next comes the brick, nine-span Frampton Mansell Viaduct[3] and the line then follows the Golden Valley down to Stroud, passing through Chalford Gorge, where railway, A419 road, River Frome and the Stroudwater and Thames and Severn Canals run close together. After Stroud, a sharpish curve northward out of the valley through Stonehouse leads to the junction made on 12 May 1845 with the Bristol and Gloucester line at Standish. About 1½ miles beyond the junction an accommodation track and public footpath are carried over the line by an interesting cast-iron bridge.[4] The three open-spandrel arch ribs, springing from brick abutments, each consist of two half-span castings, bolted together at the crown.

The Bristol and Gloucester Railway[5,6] began as the Bristol and Gloucestershire Railway, 6 miles of standard gauge horse-drawn tramroad from the River Avon, opened in 1835 to serve the north Bristol coalfield.

Under an Act of 1839, an extension, with steam traction, was authorized to Standish Junction and the name was changed to the Bristol and Gloucester Railway, with broad gauge tracks and running powers over the line from Standish to Gloucester. At Gloucester the broad gauge lines met the standard gauge Birmingham and Gloucester Railway. The need to transfer goods as well as passengers at the break of gauge soon demonstrated in a practical fashion what had hitherto been a matter for theoretical discussion, namely the disadvantage of having more than one gauge in the national railway system.

In 1845 the Midland Railway acquired both the Birmingham and Gloucester and the Bristol and Gloucester lines and by 1854 had added standard gauge throughout the latter.

An interesting sidelight on the battle of the gauges is that in 1832 the Kennet and Avon Canal Company opened the Avon and Gloucestershire Railway,[7] a standard gauge horse tramway from near Keynsham to a junction with the Bristol and Gloucestershire at Mangotsfield, with running powers over a further 2½ miles northward. Engineers of the new Bristol and Gloucester had therefore from the outset to lay standard gauge track over this length to accommodate the Avon and Gloucestershire and so established the first example of mixed gauge track in Britain.

Over the route so far described the *Cheltenham Flyer* was, in the 1930s, making the headlines with daily fast runs to London and back.

The line to Gloucester leaves the Gloucester to Birmingham line at Tuffley. The first 7½ miles after Gloucester, to Grange Court, were Great Western from the outset; beyond that the line is that of the South Wales Railway.

To Chepstow the right bank of the Severn is followed, past the site of the Severn Railway Bridge and across Lydney Docks. After Chepstow the line passes under the M48 motorway just after the latter comes off the Wye and Severn bridges. It passes Portskewett, at the western end of the Bristol and South Wales Union Railway and the Sudbrook Pumping Station of the Severn Tunnel, and near the tunnel entrance crosses the main line from London and Bristol, which joins it at Severn Tunnel Junction.

The South Wales Railway opened from Chepstow to Swansea in 1850 and from Grange Court to Chepstow East in 1851, but the final link across the Wye at Chepstow, Brunel's iron bridge, was not completed until 1852. The design of the 300 ft main span, with suspension chains tied by tubular arched girders, was not unlike that of his later famous Saltash Bridge.[8] The plate girder approach spans were replaced in 1948 and in 1962 the successors of the firm which built the original bridge replaced the main span by welded underline Warren girders.[9]

Although the branch lines such as those to Tetbury and Cirencester (where the station building is incorporated in the town's bus station) and the several feeder lines from the Forest of Dean collieries have long since closed, the route remains well used as part of the national rail network.

1. MacDermot E. T. (revised Clinker C. R.) *History of the Great Western Railway Volume 1 1833-1863.* Ian Allan, London, 1982, 79–85.

2. Sapperton Railway Tunnels (HEW 1211) SO 940 022.

3. Frampton Mansell Viaduct (HEW 389) SO 919 028.

4. Stonehouse Cast Iron Bridge (HEW 935) SO 810 090.

5. Maggs C. G. *The Bristol and Gloucester and Avon and Gloucestershire Railways.* Oakwood Press, Lingfield, 1969.

6. Bristol and Gloucester Railway (HEW 1075).

7. Avon and Gloucester Railway (HEW 1074).

8. Otter R. A. *Civil engineering heritage: Southern England.* Thomas Telford, London, 1994, 41–42 (HEW 29).

9. Berridge P. S. A. *The girder bridge.* Maxwell, London, 1969, 82–136.

11. Town Bridge, Lechlade

HEW 628
SU 214 993

The A361 road running south from Lechlade is carried over the River Thames by Town Bridge, also known as Halfpenny Bridge because of the toll levied on foot passengers until 1869.

This well-proportioned bridge[1] set in pleasant surroundings was built in 1793 by the Burford and Swindon Turnpike Trust. The designer was Daniel Harris, one-time Engineer to the Thames Commission. James Hollinsworth, a mason from Banbury who had worked on the Oxford Canal, was the contractor.

A 51 ft span Cotswold stone arch, rise 15 ft, is flanked on each side by a 10 ft span flood arch, plus a 6 ft span towpath arch on the south bank. The width between parapets is 20 ft with a roadway width of 12 ft.

The bridge is unusual in this part of the country on two counts. First, the masonry courses in the spandrel walls are laid radially rather than as conventional horizontal courses. There are other good examples of this style of masonry in the skew arch at Rainhill, Lancashire[2] and at Wicker Arches, Sheffield.[3] The second special feature is

Town Bridge, Lechlade

that the toll-house is built integrally with the north abutment.

The bridge was extensively repaired in 1958 and strengthened in 1973, without altering its external appearance.

1. PHILLIPS G. *Thames crossings. Bridges, tunnels and ferries.* David and Charles, Newton Abbot, 1981, 13–16, 19.

2. RENNISON R. W. *Civil engineering heritage: Northern England.* Thomas Telford, London, 1996, 250–251 (HEW 553).

3. Ibid. 200–201 (HEW 498).

12. Whitney Toll Bridge

HEW 816
SO 259 474

Whitney Toll Bridge Act of 1797 authorized the building of a bridge in timber and stone over the River Wye at Whitney, near Hay-on-Wye. The present bridge, built under this Act and completed about 1802, is not the first bridge on the site. The earliest bridge, built about 1774, was—like two subsequent bridges—destroyed by floods; the third bridge, a stone structure with five arches, fell in the great flood of February 1795.[1,2]

The 1802 bridge is unusual, consisting of a stone arch at each end with a centre section 110 ft long built of

R. CRAGG

Whitney Toll Bridge

timber. The timber section is in three roughly equal spans, supported on two timber piers, themselves founded on masonry sub-piers. There are diagonal timbers running from the piers to the bridge beams to give additional support. The roadway is constructed of macadam surfacing laid on a timber deck and is 14 ft wide. The bridge carries a 10 ton weight restriction.

A small toll-house is situated at the north end of the bridge and is in use, the toll rights being privately owned. The bridge stands just south of the junction between the B4350, which crosses it, and the A438.

1. JERVOISE E. *The ancient bridges of Wales and Western England*. Architectural Press, 1936; republished by E.P. Publishing, East Ardsley, Wakefield, 1976, 114.

2. GORVETT D. *Bridge over the River Wye*. C. J. and A. Bryant, Whitney, 1984.

13. Bredwardine Bridge

HEW 866
SO 336 447

Bredwardine Bridge, considered by Jervoise to be 'one of the finest brick bridges in England',[1] carries a minor road over the River Wye near the Herefordshire village of Bredwardine, and has stood almost unchanged since it was built in 1769 to replace two ferries.

R. CRAGG

Bredwardine Bridge

Six semicircular arches of about 32 ft span, supported on brick piers, carry a 13 ft roadway, the total length of the bridge being 285 ft. The piers have triangular cut-waters both upstream and downstream and on two of the piers they are continued upwards to form pedestrian refuges in the parapets. Only the four middle arches are over the river, the other two serving as flood relief spans.

It was originally a toll bridge. Tolls were removed in 1824 but reimposed in 1863 to help the trustees with the cost of repairs. In 1894 Hereford County Council assumed responsibility for the bridge and in 1922 carried out major repairs to the foundations and arches.

1. JERVOISE E. *The ancient bridges of Wales and Western England.* Architectural Press, 1936; republished by E.P. Publishing, East Ardsley, Wakefield, 1976, 114–115.

14. Wye Bridge, Hereford

HEW 1879
SO 508 396

The six-span masonry arch bridge crossing the River Wye in the heart of Hereford is substantially as originally set out when it was rebuilt in 1490, albeit with considerable alterations carried out in the seventeenth and nineteenth centuries.[1]

The bridge is about 250 ft long between the river banks and comprises six arches on five river piers. The piers are approximately parallel to the river flow and slightly angled to the arches. The bridge is built of local red and buff sandstone faced with ashlar masonry. The spans of the arches vary but are generally between 28 ft and 31 ft, the width of the piers being between 14 ft and 16 ft. Numbering from the north (city) side, arches 1, 2, 4 and 5 are late fifteenth-century four-centred arches. Many changes in detail have taken place over the years but typically, arch number 4 has a span of about 30 ft 8 in., rises 13 ft from its springing and the intrados is 18 ft above low water. The old arches are set back and those in bays 4 and 5 are partly obscured by widening carried out in 1826. Traces of earlier work in the form of arch ribs or a lower curve of intrados are visible below the springing in bays 2, 3 and 4; these may be part of the pre-1490 structure.

Arch number 3 was demolished in 1645 and rebuilt shortly afterwards with a segmental section and three shallow ribs. The southernmost arch, number 6, has a plain seventeenth-century, or earlier, arch supported by a later segmental arch. Further arches of higher profile were added to both sides of the bridge in 1826 to enable widening of about 3 ft to be carried out on either side.

Wye Bridge, Hereford

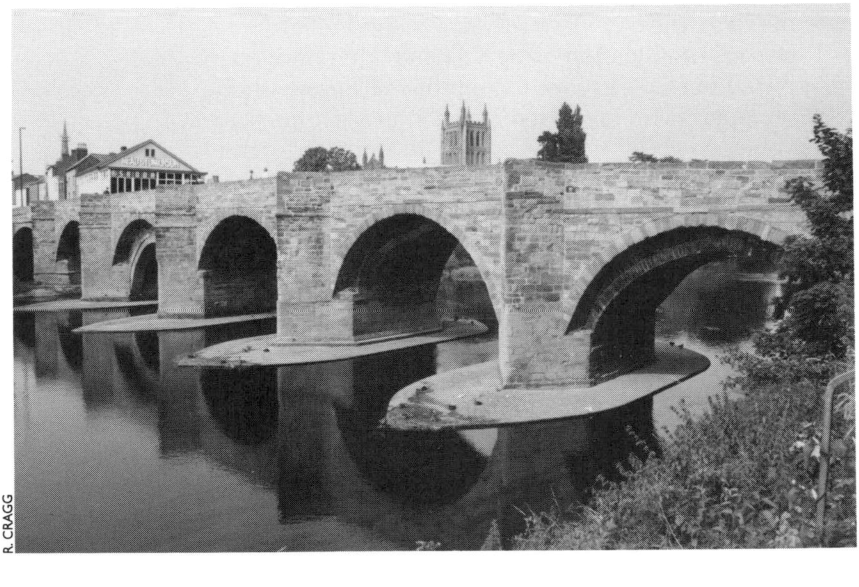

R. CRAGG

1. JERVOISE E. *The ancient bridges of Wales and Western England.* Architectural Press, 1936; republished by E.P. Publishing, East Ardsley, Wakefield, 1976, 115–116.

15. Quay Pit Bridge, Tewkesbury

HEW 387
SO 892 330

At the end of a short side-turning off the A38 in the middle of Tewkesbury, a 52 ft span bridge crosses the River Mill Avon just above the locks joining that river to the Severn.

Five cast-iron arched ribs with a rise of about 11 ft spring from water level, their spandrels infilled with circular hoops. The deck plates are cast iron and the outer ribs are surmounted by solid cast-iron kerbs which carry cast-iron railing fences, with ornamental panels at mid-span. The bridge is thought to have been built in 1822 but may be earlier.

In recent years the bridge was used by heavy lorries to and from the flour mill across the river. The deck level was raised by adding metalling and bitumen macadam, and additional kerbs, limiting the road width to 19 ft 6 in., were laid to keep vehicles from contact with the fencing. As a result of concerns about its structural capacity, heavy traffic was diverted in 1996 onto a nearby disused railway bridge.

16. Mythe Bridge, Tewkesbury

HEW 134
SO 889 337

This structure[1] is described in an original communication of 11 March 1828 to the Institution of Civil Engineers by its Engineer, Thomas Telford himself, who considered the bridge to be rather special.[2] The Resident Engineer was William Mackenzie until his departure for a similar position on the Birmingham Canal.

It is the largest of several similar bridges cast by William Hazledine of Shrewsbury and has a main span of 170 ft, consisting of six cast-iron arch ribs, each of eight segments of five panels of open web X type. The spandrels also have open X pattern bracing with cross bolts and distance pieces at the intersections. The deck is of cast-iron flanged plates; the ballast plates are solid panels with the date 1825 cast on the centre ones and there is a vertical paled parapet with stone pilasters. The width

IRONBRIDGE GORGE MUSEUM TRUST

Mythe Bridge

between the parapets is 24 ft. The very substantial abutments are piled and incorporate six tall Gothic pointed arches of 12 ft span on each side.

The bridge carries the A438 Ledbury road over the River Severn just west of Tewkesbury. A 7.5 ton weight limit, accompanied by restriction to single-line traffic, was imposed following a comprehensive assessment of its structural capacity in 1990. During 1992 overstressed compressive members were strengthened with steel plates, which were bonded to the cast iron using epoxy resin, and additional vertical and plan bracings were installed. As a result, the weight limit was raised to a more tolerable 17 tons.

1. MACKENZIE W. Account of the bridge over the Severn, near the town of Tewkesbury. *Trans. Instn Civ. Engrs*, 1838, **2**, 1–14.

2. TELFORD T. *Account of Tewkesbury bridge.* Unpublished manuscript, Institution of Civil Engineers (Original communication No. 126, 11 March 1828).

17. Great Malvern Station

HEW 1113
SO 784 457

This station, opened in 1863 on the line between Worcester and Hereford, was provided by its architect, E. W. Elmslie, with a magnificent cast-iron platform roof.

Column head, Great Malvern Station

Cast-iron columns support decorated beams and cantilevers running parallel and perpendicular to the platform edge. There is a pitched glass roof and a timber canopy.

Of the 28 columns (14 on each platform), 26 are intricately decorated at the column heads by ornate castings representing the leaves and fruit of trees. There is a large variety of different designs and all are brightly painted.

The station was extensively restored in 1987–88 and the majority of the station buildings are now used for other purposes including Lady Foley's Tearooms, a restaurant and offices.

Railway stations are on the face of it purely architectural works. However, they have a wider function in that they are the point of contact between the public and the railway, in which respect they compare with seaports

R. CRAGG

Powick Iron Bridge

and airports. Moreover the platforms, track layouts and often the ancillary buildings are essentially engineering works.

18. Powick Bridges, Worcester

HEW 1620
SO 836 524

Crossing the River Teme just south of Worcester are two interesting bridges.

The older of the two is now bypassed by the younger and is a three-span stone bridge over the river with a further two spans, one in brick, over a former mill tail race immediately to the north. The date of the bridge is unknown but it is referred to in records of 1447, 1598 and 1599 concerning its upkeep. The arches are segmental with spans of about 20 ft over the river and 25 ft over the mill race, all the piers having substantial cutwaters on both sides which are extended upwards to form pedestrian refuges. The main river arches have a substantial skew angle of 30°.[1]

To the east of the medieval bridge is its replacement, a fine cast-iron single main arch with two small cast-iron side arches, which now carries the main road across the river. The Engineer for the new bridge was C. H. Capper of Birmingham and it was completed in 1837. The main arch has a span of 68 ft 8 in. and is segmental in form with a rise of 9 ft 11 in. Seven ribs each formed from three segments with bolted joints are braced laterally by diaphragms at each of the two segment joints and by 16 tie rods. The spandrel has an X lattice which supports the roadway. The deck was originally constructed from cast-iron plates; a modern reinforced concrete deck slab was cast in 1957 and 1968 using the plates as permanent formwork. The sandstone abutments carry decorative pillars at each corner of the main span and a large decorative string course below the springing level of the arch. The two side arches, for flood relief, have pointed arches of 15 ft span with seven ribs. The ribs are cast in one piece with a single X lattice in the spandrel. There is a nice cast-iron parapet rail over the main span with a pillar at centre span bearing a coat of arms below.

1. JERVOISE E. *The ancient bridges of Wales and Western England*. Architectural Press, 1936; republished by E.P. Publishing, East Ardsley, Wakefield, 1976, 150–151.

19. Holt Fleet Bridge

HEW 135
SO 824 634

Holt Fleet Bridge, built in 1828, carries the A4133 road across the River Severn in a single span of 150 ft and was designed by Thomas Telford. As originally built, the arch

Holt Fleet Bridge

R. CRAGG

was made up of five cast-iron ribs, each 3 ft 2 in. deep, made in seven segments with Telford's familiar open X web. The deck is supported from the arch by cast-iron struts inclined to the vertical and arranged in intersecting pairs in the form of an X. These struts are cross-connected at the point of intersection by tie rods with distance pieces. They are secured by mortise and tenon joints with cast-iron wedges. At deck level, cast-iron beams 6 in. deep and 2 in. wide carried the deck plates and the roadway. The massive sandstone abutments of the bridge are pierced on both sides of the river by flood arches.

In 1928 the bridge was extensively strengthened by encasing the upper and lower edges of the arch ribs in reinforced concrete extending across the whole width of the bridge. In addition, one of each pair of spandrel struts was also encased in concrete. A new reinforced concrete deck was cast on new cross beams and the roadway was widened by cantilevering out over the sides of the old arch.[1]

1. HAMMOND B. C. The strengthening of a cast iron bridge by welded steel bars encased in concrete. *Sel. Engng Papers Instn Civ. Engrs*, 1934, No. 162, 3–15.

20. Birmingham and Gloucester Railway

HEW 1091
SP 070 867 to
SO 838 185

In 1832, Brunel made a survey for a broad gauge railway to link Birmingham and Gloucester, but his route was rejected by the promoters in favour of an alternative standard gauge route, proposed the following year by Captain W. S. Moorsom. The Birmingham and Gloucester company was incorporated in 1836. The route adopted was from Birmingham to Cheltenham, where it made use of the line of the existing Gloucester and Cheltenham tramway to gain access to Gloucester.[1]

Construction started at Cheltenham, working northwards towards Birmingham, and the line was opened in stages, reaching Camp Hill (a temporary terminus in Birmingham) in December 1840. The railway was empowered by its Act of Parliament to use the terminal station of the London and Birmingham Railway (L&BR) at Curzon Street, to which it gained access via Gloucester

Junction, running west over L&BR tracks to Curzon Street. The final link was opened on 17 August 1841.

In 1854 the Birmingham and Gloucester (B&G) trains were transferred to the newly built station at New Street. In 1864 an end-on junction was made between the B&G and the Birmingham and Derby Junction Railway,[2] thus enabling through running between Derby and Gloucester without the need to reverse in New Street.

HEW 865
SO 969 693 to SO 992 720

The route of the B&G Railway follows the Severn Valley northwards from Cheltenham for most of its length, but at Bromsgrove, the railway is faced with a steep ascent up to the Birmingham area over the shoulder of the Lickey Hills. The great Lickey Incline was built 2 miles long at a gradient of 1 in 37.7 (2.7 per cent) to lift the line over this obstacle. Opened in September 1840, the incline is straight with double tracks. It starts immediately north of Bromsgrove station and climbs to the summit near Blackwall, a vertical height of about 300 ft.

The incline was designed from the outset to be used by locomotive-hauled trains. In the early days of its operation special locomotives, built by Norris of Philadelphia, were imported from the United States of America, but these were replaced in 1845 by a more powerful locomotive designed by James McConnell, the Locomotive Superintendent of the B&G, and by others in succeeding years. As the weights of trains increased, the need for an even more powerful banking engine became more urgent and the Midland Railway (which had taken over the Birmingham and Gloucester in 1846) introduced a unique 0–10–0 locomotive designed by Sir Henry Fowler, which was in use from 1920 until 1956. Modern diesel locomotives normally haul passenger trains up the incline unaided; banking assistance can be provided for freight trains if necessary.

In the churchyard at Bromsgrove are the graves of two employees of the B&G who lost their lives in an accident on the Lickey Incline in 1840. The tombstones depict locomotives of the period.

Today the Birmingham and Gloucester line is an important part of the north-east to south-west route in the national railway network.

1. LONG P.J . and AWDRY Revd W. V. *The Birmingham and Gloucester Railway*. Alan Sutton, Gloucester, 1987.

2. Birmingham and Derby Junction Railway (HEW 1667).

21. Tardebigge Locks

HEW 768
SO 963 680 to SO 995 693

Long before the coming of the Birmingham and Gloucester Railway and its Lickey Incline, the builders of the Worcester and Birmingham Canal had also faced the steep climb out of the Severn Valley to the higher ground around Birmingham.

At Tardebigge, between Redditch and Bromsgrove, they built the longest flight of locks in Britain, 30 in 2½ miles, giving a total rise of 217 ft. These locks are numbered 29–58: they are closely preceded by another six at Stoke Prior, numbered 23–28, with a rise of 42 ft, and by yet another six at Astwood, numbered 17–22 and another 42 ft rise. Thus the total rise of the canal in only 5¼ miles is 301 ft.

The lock chambers are 74 ft 6 in. long and 7 ft 4 in. wide, to take narrow boats not exceeding 71 ft 6 in. long and 7 ft beam. They have single leaf top gates and double mitred

Tardebigge Locks: top lock, site of the experimental lift

R. CRAGG

bottom gates and the average distance between the locks is 250 to 300 ft.

The rise of each lock is about 7 ft except for the Top Lock which has a rise of 14 ft. This lock chamber was originally the site of trials carried out between 1808 and 1813 of a vertical lift designed by John Woodhouse. The lift consisted of a tank suspended by chains which passed over large wheels, the weight of the tank being counterbalanced by brick weights. The tank weighed 64 tons when filled with water, and was raised and lowered by hand winches. Gates at each end of the tank allowed boats to enter and leave. Although some successful operation of the lift was obtained, the mechanism was too complex for the technology of the time and the lift was replaced by a conventional lock. The whole canal was finally opened in December 1815.[1]

The Engineers for the Worcester and Birmingham Canal were Thomas Cartwright (until July 1809), John Woodhouse (July 1809 to mid-1811) and William Crosley (from mid-1811).

1. Hadfield C. *The canals of the West Midlands.* David and Charles, Newton Abbot, 1985, 140–142.

22. Stourport-on-Severn Canal Terminus

HEW 241
SO 811 711

James Brindley sited the terminal basins of the Staffordshire and Worcestershire Canal (see Chapter 6) on a terrace of land about 30 ft above the River Severn, well above any danger from floods. Originally, the canal entered the Upper Basin (the more easterly of the high-level basins) which communicated with the river through wide barge locks. This development was opened in 1770. A new basin to the west of the Upper Basin, and connected to the river by a set of narrow locks, opened in 1781.[1]

Between these two basins is the Clock Warehouse, prominent among the original canal buildings, which also include the Tontine Inn, dating from the late eighteenth and early nineteenth centuries.

The Clock Warehouse is a red-brick two-storey building, 126 ft long and 25 ft wide with a slate roof. At the

R. CRAGG

Stourport Canal Basin with the Clock Warehouse

south-west corner a single-storey lean-to has been built on, evidently at a later date as its roof cuts the first floor window openings. At the centre of the roof ridge a decorative timber cupola houses the four-faced clock which gives the warehouse its name.

The precise date of construction of the building is not known but it is recorded that the clock was installed in 1812. Originally used as a store for grain and general goods, and later as part of a timber yard, it now houses the headquarters of a local yacht club.

1. LANGFORD J. I. *A towpath guide to the Staffordshire and Worcestershire Canal*. Goose and Son, Cambridge, 1974, 176–188.

23. Stanford Bridge

HEW 1619
SO 714 658

This single-span reinforced concrete arch bridge, built in 1905, is probably the oldest open-spandrel reinforced concrete bridge in Britain. It is the fourth bridge to span the River Teme at this point. The first bridge, a timber structure dating from 1548, was replaced by a three-arch stone bridge which, in turn, was replaced in 1797 by a single-span iron bridge.[1]

The bridge has a single span of 96 ft and the segmental

R. CRAGG

Stanford Bridge

arch rises about 13 ft. There are three arch ribs, each 2 ft 7 in. deep and 1 ft wide, and the bridge deck is supported by three spandrel struts on each side. Lateral stiffness is provided by diaphragm walls which connect the arch ribs. The overall width of the bridge is 10 ft 2 in., the deck slab being extended beyond the sides of the arch. The arch springs from stone abutments with brick buttressing which appear to be relics of the previous iron bridge. Just to the north of the bridge, three red-brick segmental flood arches of 10 ft span pierce the approach embankment.

A plaque on the bridge records that it was rebuilt by the Worcestershire County Council in 1905. It was designed by L. G. Mouchel and Partners.

As the roadway over the bridge is only 10 ft 2 in. wide, with two footways, it has proved to be inadequate for modern traffic. A new concrete bridge has been built about 200 yd downstream from the old bridge and this interesting structure is now preserved as a footbridge.

1. JERVOISE E. *The ancient bridges of Wales and Western England*. Architectural Press, 1936; republished by E.P.Publishing, East Ardsley, Wakefield, 1976, 149–150.

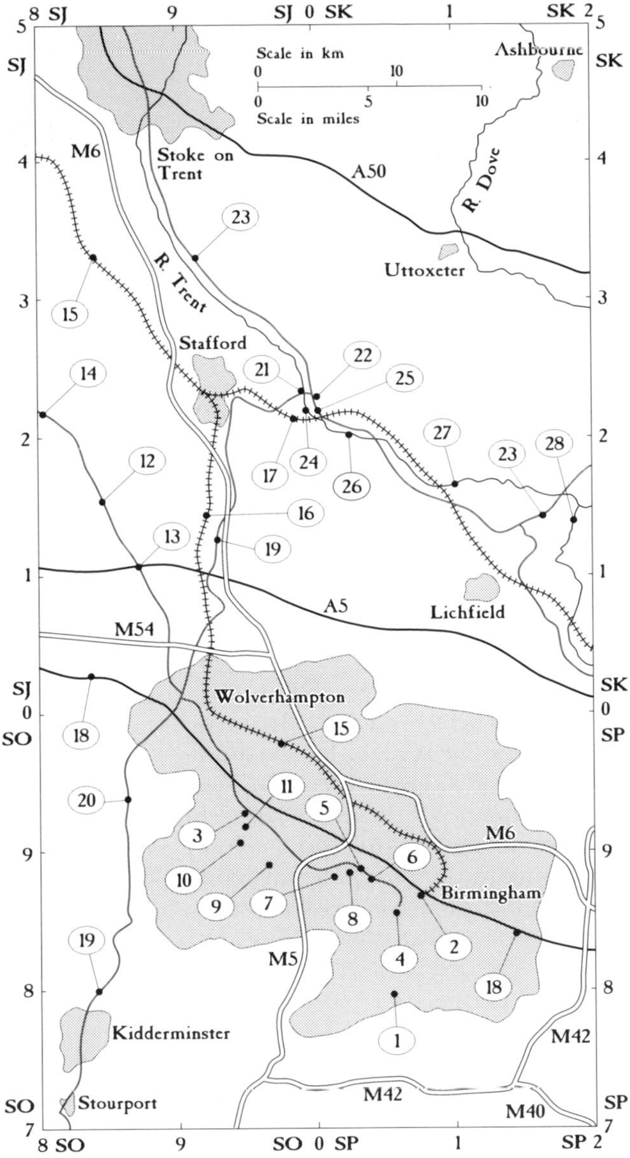

1 King's Norton Stop Lock
2 Curzon Street Railway Station, Birmingham
3 The Birmingham Canal Navigations
4 Gas Street Basin
5 Engine Arm Aqueduct, Smethwick
6 Rabone Lane Canal Junction Bridges
7 Smethwick Cutting
8 Galton Bridge
9 Netherton Canal Tunnel
10 Dudley Canal Tunnel
11 Tipton Lift Bridge
12 Birmingham and Liverpool Junction Canal
13 Stretton Aqueduct
14 Shelmore Bank
15 The Grand Junction Railway
16 Penkridge Viaduct
17 Shugborough Tunnel
18 The Holyhead Road
19 Staffordshire and Worcestershire Canal
20 Bratch Locks
21 Trent Aqueduct, Great Haywood
22 Great Haywood Canal Junction Bridge
23 Trent and Mersey Canal
24 Shugborough Hall Bridges
25 Essex Bridge, Great Haywood
26 Wolseley Bridge
27 Mavesyn Ridware Bridge
28 Chetwynd Bridge, Alrewas

6. West Midlands and South Staffordshire

Geographically, much of this region lies squarely on the watershed of England and thus there are few large natural rivers. In the north rises the high ground of Cannock Chase, while to the south the Birmingham plateau is bordered by the Clent and Lickey hills before the descent into the valley of the River Severn and its tributaries.

Economically it is a region dominated by the intensively developed area between Birmingham and Wolverhampton known colloquially as 'The Black Country'. Developing rapidly in the late eighteenth and nineteenth centuries, this area became a major centre of industry. Its growth was mainly fuelled by the metal trades: nail making and chain making, locksmithing and gunsmithing and the production of brassware, screws, nuts and bolts together with heavy engineering of high quality. Along with industry developed a demand for transport—always a necessary adjunct to the growth of factory-based manufacture, which needs to convey raw materials to the manufactory and finished goods to the markets. Partly because of the lack of suitable natural waterways, canals were developed in the area to an extent not seen elsewhere and much of the network remains in use today. Birmingham was also served by two of the first major trunk railways, the London and Birmingham and the Grand Junction, and much early railway development took place here.

Although roads did not play a major part in the early development of the region, a notable exception is the Holyhead Road, improved under the guidance of Thomas Telford in the early nineteenth century, which passes through the area by way of Birmingham and Wolverhampton. More recently, the area has been the scene of much road building, being traversed by the M5, M6 and M42 motorways with spectacular junctions at Gravelly Hill and Ray Hall.

Many of the early civil engineers were active in the provision of transport facilities in the region and some familiar names will be found in the following pages: Thomas Telford, James Brindley, Robert Stephenson and Joseph Locke, to name but a few. The well-ope contractor, Thomas Brassey, started his railway building career here with the construction of Penkridge Viaduct on the Grand Junction Railway.

To the north of the region a more rural landscape predominates in contrast with the 'dark satanic mills' of the Black Country. Even here the civil engineer has left his mark with the Trent and Mersey and Staffordshire and Worcestershire Canals, and the Grand Junction Railway and Trent Valley line. These formed important links with the developing industrial areas of the north of England.

1. King's Norton Stop Lock

HEW 975
SP 056 795

The Stratford-upon-Avon Canal was authorized by an Act of Parliament of 1793 to run from a junction with the Worcester and Birmingham Canal at King's Norton to Stratford-upon-Avon. As the level of water in the Stratford Canal was higher than the level in the Worcester and Birmingham Canal, a stop lock was built at King's Norton to prevent uncontrolled transfer of water from one canal to the other.

R. CRAGG

King's Norton Stop Lock

The lock is remarkable in having vertical 'guillotine' gates instead of the usual vertically pivoted leaf lock gates. They are built in timber with a metal frame and are raised and lowered in a cast-iron guide frame. The movement of each gate is controlled by a winch sited on the north side of the lock. A lifting chain runs over a large pulley wheel mounted on the gate framework and round a single sheave pulley attached to the top of the gate. A counterbalance weight is suspended in a 'well' on the south side of the lock and is attached to the top of the gate by a chain, which runs over two large pulley wheels. One of the advantages of the guillotine gate where the water level difference is small is that it removes the need for paddle gear.

The water level in the Worcester and Birmingham Canal was eventually raised to that of the Stratford Canal, thus making the lock unnecessary, so both gates are now kept permanently raised to permit the uninterrupted passage of boats.

Another unusual feature of King's Norton Lock is that the lock chamber is crossed by a road bridge. This bridge has recently been rebuilt and provides an excellent example of skew arch brickwork.

The Engineer to the Canal Company at the time of its construction was Josiah Clowes.

2. Curzon Street Railway Station, Birmingham

HEW 420
SP 078 871

Robert Stephenson's standard gauge London and Birmingham Railway was, at 112 miles, the longest continuous railway then built and is generally regarded, together with the Stockton and Darlington,[1] the Liverpool and Manchester,[2] and the Grand Junction railways, as being one of the four pioneering real public railways. The London and Birmingham reached Birmingham on 24 June 1838, the whole line from London having involved heavy engineering works through difficult country. Stephenson's assistant for the section from Rugby to Birmingham had been Thomas Gooch, brother of Brunel's Locomotive Superintendent on the Great Western Railway.

The terminus in Birmingham was not at the present

R. CRAGG

New Street station, but about half a mile to the north-east at the junction of Curzon Street and New Canal Street.

Curzon Street Station

The station, opened in 1842, was designed by Philip Hardwick, architect of the original Euston station, now vanished together with its great Doric arch, but some features of it are reflected in its counterpart at the Birmingham end of the line. The squarish three-storey main station block which remains today is fronted by four giant Ionic columns which were presumably intended to contrast with the Doric order at Euston. The first floor windows are balustraded. The rear of the building has two further columns between outer square pilasters. The inside contained what was originally the booking hall with a steep iron balustraded stone staircase, a refreshment room and offices. The station was built at a cost of £26 000. An early engraving by Bourne of the station shows the main building flanked by two archways leading into the station, but recent excavations have not revealed any trace of these structures and it is suggested that although they were included in Hardwick's designs they were never built.

The station was known simply as 'Birmingham' until November 1852, when the suffix 'Curzon Street' was adopted to distinguish it from New Street and the newly opened Great Western Railway station at Snow Hill. In July 1854 all regular trains were transferred to New Street which was reached by a short line, also engineered by Robert Stephenson, from near Curzon Street.

Thereafter, apart from holiday excursion trains which ran until 1893, Curzon Street handled goods traffic only. The building has recently been completely refurbished and it is now used as the headquarters of the Prince's Trust.

It was at the Queen's Hotel, adjacent to the station building but now demolished, that the Institution of Mechanical Engineers was founded on 27 January 1847 with George Stephenson as its first President. The plaque commemorating this event is now displayed in the station building.

1. RENNISON R. W. *Civil engineering heritage: Northern England*. Thomas Telford, London, 1996, 83–84 (HEW 85).

2. Ibid. 249–255 (HEW 223).

3. Birmingham Canal Navigations

HEW 1182

No review of the civil engineering heritage of Britain would be complete without a reference to the canal system which developed in Birmingham and the Black Country between 1768 and the 1830s. In this area the use of the canal for commercial transport was developed to a degree unsurpassed anywhere in the country.[1]

The first canal in the area, built by the Birmingham Canal Company and opened in 1772, ran 22½ miles from a junction with the Staffordshire and Worcestershire Canal at Aldersley, west of Wolverhampton, to wharves in what is now central Birmingham near to Broad Street and Summer Row. It ran at three distinct 'levels': 453 ft above sea level from Birmingham to Smethwick, 491 ft over the high ground at Smethwick and 473 ft from Smethwick to Wolverhampton. The levels were connected by flights of locks and at Wolverhampton the canal descended to its junction with the Staffordshire and Worcestershire Canal by a flight of 20 locks (later increased to 21). The Engineer

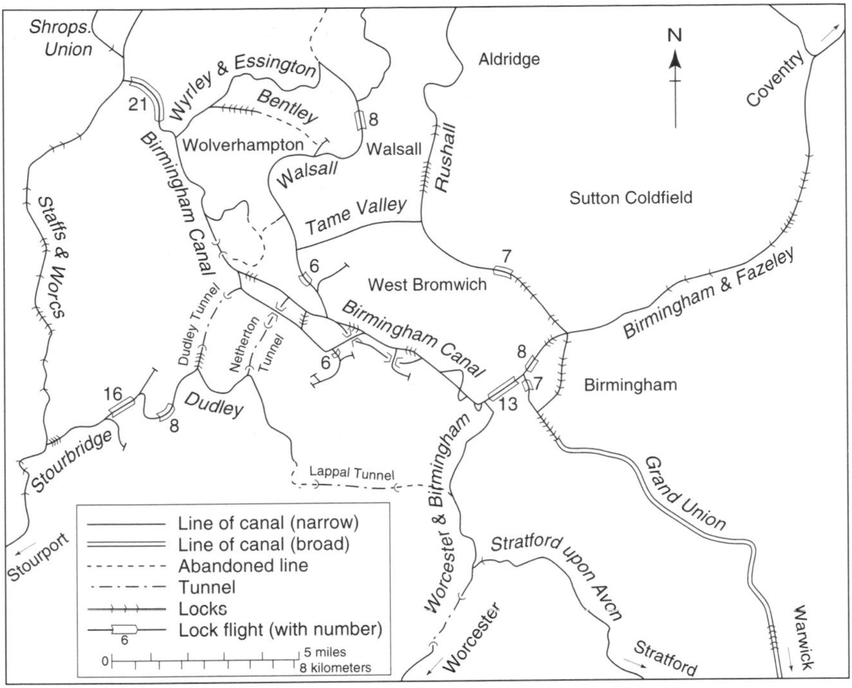

Birmingham Canal Navigations

for the project was the pioneering canal engineer James Brindley.

Later, major improvements were carried out along the line of this canal, notably the progressive lowering of the awkward summit level at Smethwick, first to 473 ft in 1789 and subsequently, by means of Thomas Telford's great cutting, to 453 ft in 1829. In 1837 the Deepfields to Tipton cut-off through Coseley Tunnel was opened and in 1838 a new line of canal between Smethwick and Tipton was constructed.

The Birmingham Canal formed the focus of an intensive development of 'narrow' canals promoted by the Birmingham Canal Company and other companies, which included the Dudley Canal, the Birmingham and Fazeley Canal, the Wyrley and Essington Canal, the Tame Valley Canal, the Rushall Canal and others. In addition, a large number of branch canals were built, serving the industry which grew up along their banks. Most of the canals served local needs, but the Birmingham network was connected to the 'national' canal system by the Staffordshire and Worcestershire, the Warwick and Birmingham (later part of the Grand Union), the Stratford-upon-Avon, the Worcester and Birmingham and the Coventry canals. In total, over 200 miles of canal were built in the area, more than in the city of Venice.

Water pumped from the collieries was an important source in parts of the network and a complex system of water management was developed with many pumping stations and reservoirs. Some of the old system has been abandoned and left to decay but much still survives and remains in use. Although subject to railway competition, the canals continued to act as a local distribution network with many railway to canal interchange basins being established; it was road-based competition which finally killed off commercial traffic on the Birmingham canals.

Recently, much effort has been made to improve the canal-side environment in central Birmingham,[2] especially with the development of the new National Convention Centre and National Indoor Arena alongside the canal just to the north of Gas Street Basin.

1. BROADBRIDGE S. R. *The Birmingham Canal Navigations, Volume 1: 1768–1846*. David and Charles, Newton Abbot, 1974.

2. 'Cut and dried.' *New Civil Engineer*, 1995, 26 Jan., 40.

4. Gas Street Basin

HEW 1528
SP 065 864

Gas Street Basin has been described as 'the hub of Britain's canal system'. It lies at the heart of the great network of canals of Birmingham and the Black Country serving the industries which grew up along their banks in the second half of the eighteenth century.

The Basin is all that is left of a complex system of basins and wharves which formed the terminus of the original section of the Birmingham Canal when it was opened in 1772.[1] It comprises an extensive area of water between Gas Street and Bridge Street and is flanked on the Gas Street side by a number of old canal buildings including the toll office. Originally the basin extended to the east beyond Bridge Street to extensive wharves; these are now filled in and built over, but some remains of the bridges over the canals which connected the basins can still be seen.

The basin is bisected by a narrow strip of land which originally marked its southernmost boundary. The canal and basin to the south of this land were part of the Worcester and Birmingham Canal, opened in 1795. At first the Birmingham Canal would not allow any physical junction between the two canals which were separated by this strip of land; known as the 'Worcester Bar', it was about 7 ft wide, and all goods which were to be transshipped from one canal to the other had to be carried across it.[2] By 1815 an agreement had been reached between the two companies and the Bar was breached by a stop lock, opened on 21 July of that year; the gates of this have since been removed as the canals are now at the same level. The building just to the south of the stop lock was the canal offices of the Worcester and Birmingham Canal Company.

Much restoration work has taken place in this area over the past few years and most of the east side of the basin has been redeveloped, including a new public house, appropriately named after James Brindley, the Engineer of the Birmingham Canal. The cast-iron bridge spanning the old stop lock in the Bar is of recent construction, replacing a plank swing bridge which formerly spanned the lock chamber.

Just to the north of Gas Street Basin the canal passes through a short tunnel under Broad Street which, in its varying cross-section, illustrates the various lengthenings of the tunnel which have been required by corresponding widening of the road above.

1. Broadbridge S. R. *The Birmingham Canal Navigations, Volume 1: 1768–1846.* David and Charles, Newton Abbot, 1974, 22–24.

2. Hadfield C. *The canals of the West Midlands.* David and Charles, Newton Abbot, 1985, 85–86.

5. Engine Arm Aqueduct, Smethwick

HEW 492
SP 024 888

This cast-iron aqueduct, also known as Engine Branch Aqueduct, carries the Rotton Park feeder at the 473 ft level over the later, lower line constructed in 1829. The difference between the water levels in the two canals is 20 ft.

R. CRAGG

Engine Arm Aqueduct

Designed by Thomas Telford, with the ironwork cast by the local Horseley Company, the structure consists of a cast-iron trough supported by a single arch of 52 ft span with five ribs, each comprising four sections with bolted joints. The spandrel bracings to the ribs are of the radial type intersected by a continuous arched member. The trough is supported on three of the ribs; the towpaths on either side are supported by cast-iron arcades of unashamedly Gothic ecclesiastical-style arches and columns.

The waterway is 8 ft wide and each towpath is 4 ft 4 in. wide, the one on the east side being paved in brickwork with raised strips to afford a better grip for horses' hooves.

6. Rabone Lane Canal Junction Bridges

HEW 1197
SP 029 890

These two bridges are typical of the cast-iron towpath bridges found in many places on the canal network of Birmingham and the Black Country. They were built in 1828 as part of Thomas Telford's improvements to the

Birmingham Canal and connect various towpaths at the junction of the old and new lines of canal.

Each bridge has a span of 52 ft 6 in. and the arch rises 9 ft. Their semi-elliptical form has an advantage over the equivalent segmental arch in this situation as it gives greater headroom near the abutments where the bridge passes over a towpath.

The two ribs of each bridge were cast by the Horseley Ironworks and form the parapets, each arch rib being in two sections with a bolted joint at the crown. They have an X-lattice structure with a decorative quatrefoil pattern below the handrail. The deck is of cast-iron plates with raised ribs 2½ in. high cast on their upper surfaces to help retain the earth filling which forms the footway.

The bridge abutments and wing walls are of brickwork with stone facings at the corners.

The ironwork of the bridges shows deep cuts caused by the abrasion of tow-ropes.

Towpath Bridges at Rabone Lane, Smethwick Junction

R. CRAGG

R. CRAGG

Smethwick Cutting: 473 ft level on the right, 453 ft level in the centre. The LNWR Stour Valley Railway is shown on the left

HEW 1875
SO 996 899 to SP 029 890

7. Smethwick Cutting

In 1824 the Birmingham Canal Company consulted Thomas Telford with a view to improving the line of its 1772 canal between Birmingham and Wolverhampton. Telford designed a number of improvement works along the canal including the driving of a new canal through the high ground at Smethwick to bypass the Smethwick Summit. The new canal was built at a level of 453 ft above sea level, 20 ft lower than the old line. With a width of 40 ft and towpaths on both sides, the new canal was an impressive addition to the canal landscape of the Midlands.

To enable the canal to be built, a massive cutting was required, 4000 yd long and up to 66 ft deep. The Resident Engineer for the work on the cutting was William Mackenzie and the contractor was T. Townshend. Luckily, a longitudinal section drawing has survived upon which Mackenzie recorded the progress of the excavation. This indicates that digging started in April 1827, and a handwritten note on the drawing records that water was first let into the new canal on the evening of Sunday 8 February 1829.

A number of bridges were built across the cutting, including two aqueducts: Steward Aqueduct,[1] which carries the old line of the canal over the low level canal, and Engine Arm Aqueduct, which carries a branch canal from the higher level. Other structures included the magnificent Galton Bridge, described below, Spon Lane Bridge (later widened in concrete) and several smaller road bridges.

At the time of its construction Smethwick Cutting was one of the greatest earthworks in the world.

1. Steward Aqueduct (HEW 1904) SP 002 898.

8. Galton Bridge

HEW 421
SP 015 893

To carry the road to Sandwell across the Smethwick Cutting, Telford designed Galton Bridge, a single cast-iron arch of 150 ft span which springs from abutments set high up on the sides of the cutting. There are six arch ribs, each made up from seven segments with bolted joints. The space between the arch and the road deck is filled with diagonal intersecting ribs. The details resemble those of several other Telford bridges including the

Galton Bridge

R. CRAGG

Mythe and Holt Fleet bridges. Galton Bridge was completed in 1829.

Named after Samuel Tertius Galton, a banker who served on the Committee of the Birmingham Canal from 1815–43,[1] Galton Bridge is rightly described as one of the seven wonders of the Birmingham canals. It has now been relieved of traffic by the construction of a new road running parallel to it. It is interesting to note that the designers of the new crossing have avoided the necessity for another large bridge by putting the canal through a concrete 'tunnel' and constructing an embankment to carry the roadway.

1. BROADBRIDGE S. R. *The Birmingham Canal Navigations, Volume 1: 1768–1846*. David and Charles, Newton Abbot, 1974, 123.

9. Netherton Canal Tunnel

HEW 669
SO 954 883 to SO 967 908

The last major canal tunnel to be built in England, the Netherton Tunnel was opened in August 1858, construction having started at the end of 1855. It was built parallel to and about 1½ miles east of Dudley Tunnel, in order to relieve congestion on the Dudley Canal. It connects the Dudley Canal to the south of the Dudley limestone ridge

South Portal, Netherton Tunnel

R. CRAGG

to the 453 ft level of the Birmingham Canal at its north end, the connecting canal passing under the 473 ft level just beyond the north portal of the tunnel.

The tunnel is 1 mile 1267 yd long and is exceptionally large, being 27 ft wide, including two towpaths, and 16 ft high above the water level. When opened it was lit by gas, which was later converted to electric lighting.

The construction cost of £302 000 compared with an estimate of £238 000, the difference being partly due to extra works necessitated by the condition of the ground through which the tunnel passes.

In 1983 the British Waterways Board replaced with concrete some 80 yd of brick invert which had heaved up to such an extent as to impede traffic,[1] and restored to the tunnel the distinction of being the longest navigable on a canal route in Britain.

1. MIDDLEBOE S. Repairers wary in Netherton heaving invert. *New Civil Engineer*, 1983, 30 June, 26–27.

10. Dudley Canal Tunnel

HEW 670
SO 933 892 to SO 947 917

Early in the canal building age, Lord Dudley and Ward built a short canal, completed in 1778, to link his underground limestone quarries at Castle Hill, Dudley, to the Birmingham Canal. Later this canal was linked to the Dudley Canal (also promoted by Lord Dudley) by means of a tunnel under the intervening ridge of high ground. After several setbacks the tunnel was finally opened in 1792.

At 1 mile 1412 yd, it was the fifth longest canal tunnel built in England. In so far as it also connected with underground mine workings, the tunnel has similarities with the Duke of Bridgewater's pioneering canal at Worsley.[1] Later, tunnels were driven from the north end of the main tunnel to gain access to other limestone workings, in particular the Wrens Nest branch tunnel, which was 1185 yd long and terminated in an underground basin.

The main tunnel is not continuous, for at its northern end it passes through two basins open to the air, the Shirts Mill and Castle Mill basins. The size of the tunnel varies, reflecting the various stages of its construction and the effects of mining subsidence. A short section at the southern end was rebuilt in 1884 to a larger size. At its deepest point the tunnel is about 200 ft below the surface. Most of the length is lined with brickwork but there are some unlined sections through harder rocks.

The tunnel was closed to traffic in 1962, but in 1973 was reopened by the Dudley Canal Trust. Unfortunately there was a serious failure of the tunnel's side wall near the south end in 1981 which led to closure of the tunnel once more, but the combined efforts of the British Waterways Board, Dudley Metropolitan Council and Dudley Canal Trust together with financial assistance from the European Development Fund, made possible a major rebuilding of the damaged section of tunnel. This was carried out between February 1991 and April 1992 by Fairclough Civil Engineering.[2] In addition, a short length of new tunnel was driven in 1984; this allowed electrically powered trip boats to gain access to the Singing Cavern, one of the old limestone workings, where an audio-visual show tells the story of limestone mining at Dudley.

1. RENNISON R. W. *Civil engineering heritage: Northern England*. Thomas Telford, London, 1996, 265–266 (HEW 976).

2. *Dudley Canal Tunnel 1792–1992*. Dudley Canal Trust, 1992.

South Portal, Dudley Tunnel

R. CRAGG

R. CRAGG

11. Tipton Lift Bridge

Tipton Lift Bridge

HEW 1225
SO 948 918

This vertical lift bridge, of unusual design, carried a road across an arm of Proof House Basin on the Birmingham Canal. It has a deck length of 27 ft 6 in. and a roadway width of 12 ft. The deck is lifted by four large chains, one being attached to each end of the two steel side plate girders. The chains pass over large wheels mounted at the top of vertical I-section steel columns and are connected to 6 ton counterbalance weights at each end of the bridge.

The drive to the chain wheels is taken from a hand winch by a complex system of vertical and horizontal shafts connected by pairs of bevelled gears and a short length of chain drive. The vertical columns are cross-connected at the top by steel lattice girders.

The bridge was fabricated by Armstrong Whitworth and Company in 1922 for the Great Western Railway. When the basin was closed in 1954, the bridge at that time was damaged and incapable of being lifted. After its removal from the basin it was eventually brought to the

Black Country Museum at Dudley, only a short distance away, and re-erected across an arm of the Museum's canal system.

12. Birmingham and Liverpool Junction Canal

HEW 1203
SJ 902 020 to SJ 626 553

The Birmingham and Liverpool Junction Canal (B&LJC) was opened in 1835 to connect the Birmingham Canal Navigations via the Staffordshire and Worcestershire Canal at Autherley to the Chester Canal north of Nantwich, a distance of 39½ miles. It provided an alternative route to the Trent and Mersey Canal for trade between the Mersey and the West Midlands. There was also a 10¼ mile branch, with 23 locks, from Norbury Junction (SJ 793 227) to Wappenshall (SJ 663 147) on the Shrewsbury Canal, and this linked the network of Shropshire canals with the Midlands.

The canal was constructed later than most. It was one of Thomas Telford's last works, and its style of construction suggests that its Engineer was endeavouring to compete with the up-and-coming railways.

The B&LJC pursues a direct course across the landscape, rejecting the 'contouring' approach of earlier canal builders and making use of deep cuttings and high embankments to give long stretches free of locks. Where locks do occur, they are generally grouped together in flights, as at Audlem (15 locks), Adderley Locks (five locks) and Tyreley Locks (five locks). On the main line there are 28 locks in all, together with a stop lock at Autherley Junction, one short tunnel and a number of aqueducts.

A number of deviations from the originally planned line had to be made during construction because of objections by landowners, most notably a bank and aqueduct at Nantwich with a revision to the planned end-on junction with the Chester Canal, and the great embankment at Shelmore Wood.[1]

The canal was built in four sections and the first contract, from Nantwich to Tyrley Locks, was started in 1827 with John Wilson as contractor. The second section, from Tyrley to Church Eaton, was built by William Provis who was also the contractor for the Newport Branch. The southernmost section of the main canal, from Church Eaton to Autherley, was built by John Wilson. The canal was largely completed by July 1833 but through traffic did not commence until March 1835, mainly as a result of problems encountered in the construction of Shelmore embankment (described elsewhere in this chapter).

In 1842 the canal company introduced steam tugs to tow boats in trains but they were considered to be uneconomic and horses were reintroduced.

Although Telford was the Engineer, William Cubitt deputized for him after February 1833 when Telford became ill. Telford died in September 1834, shortly before completion. By amalgamation, the canal became part of the Shropshire Union in 1846 (see Chapter 7). Commercial traffic ceased in the 1960s but the canal is now busy with pleasure craft and its future seems assured.

1. ROLT L. T. C. *Thomas Telford*. Penguin Books, Harmondsworth, 1979, 189–192.

13. Stretton Aqueduct

HEW 228
SJ 873 107

Constructed in 1832–33, Stretton Aqueduct carries the Birmingham and Liverpool Junction Canal in a cast-iron trough over what is now the A5 road south-west of Penkridge in Staffordshire.

The trough is formed of five sections, each 6 ft 6 in. long, bolted together and supported by six cast-iron arch ribs, each cast in two sections and joined at the centre of the arch. The clear span across the road is 30 ft, but the arch ribs are longer because the canal crosses the road at a skew angle. The structure is 21 ft wide, the waterway narrowing to 11 ft across the aqueduct. Towpaths are provided on both sides but only the one on the east side is currently in use.

On the east side is a cast-iron parapet rail 3 ft 9 in. high and each corner of the main arch is surmounted by an ornamental circular stone pillar. The abutments are in brickwork, curved in plan, and they include further stone columns matching those at the ends of the arch.

The name of the designer, Thomas Telford, and the date of construction of the aqueduct are commemorated

R. CRAGG

Stretton Aqueduct

by an inscription on the centre panel of the trough. Close inspection of this reveals that the name of the iron founder, William Hazledine of Shrewsbury, has at some time been removed from the inscription.

During 1961–62, the carriageway of the road under the aqueduct was lowered by about 4 ft to increase the headroom under the arch. Extra abutments were built for the aqueduct, with partial underpinning and support provided by an inverted portal-type structure beneath the new roadway.

14. Shelmore Bank

HEW 1224
SJ 805 215 to
SJ 793 228

During construction of the Birmingham and Liverpool Junction Canal, it was necessary to alter the line of the canal because of the opposition from Lord Anson, owner of Norbury Park, so that instead of passing at ground level through Shelmore Wood, it took a curved alignment across much lower ground to the west. This deviation necessitated the construction of a massive embankment 1900 yd long and up to 60 ft high. Major problems were encountered during the forming of the embankment, with both settlement and slipping of the tipped soil. Eventually, after six years' work (1829–35), the bank was finished at the cost of delaying the complete opening of the canal by two years. Stop gates are provided at both ends of the bank.

Today, thick vegetation covers the sides of the embankment but the flat side slopes still betray the difficulties encountered in stabilizing the fill. There are two road 'tunnels' under the bank with semicircular masonry arches springing from vertical side walls. These may be compared with the road tunnels at Beaminster and Charmouth.[1]

1. OTTER R. A. *Civil engineering heritage: Southern England.* Thomas Telford, London, 1994, 139 (HEW 899, 980).

15. Grand Junction Railway

HEW 1129
SP 078 871 to
SJ 800 402

The earliest trunk railway route,[1] the Grand Junction Railway was opened on 4 July 1837, having been authorized by an Act of Parliament in 1833.[2] The railway ran for 82½ miles from a junction with the newly opened Liverpool and Manchester Railway[3] at Newton-le-Willows near Warrington to its terminus in Birmingham at Curzon Street. It thus linked the expanding industrial areas of the Midlands to the industry and ports of the North West via Crewe and Stafford.

The line was laid out by George Stephenson and John Rastrick with Joseph Locke as assistant, but from the start of construction, Locke was effectively in complete charge of the works although he was not formally appointed Chief Engineer until August 1835. Several contractors were engaged in the building of this line, including John Stephenson, David McIntosh, James Trubshaw and Thomas Townshend. It was on this line that the contractor Thomas Brassey began his long association with railway construction.

The route[4] of the railway across the area covered by this chapter commenced at the station in Curzon Street, Birmingham, and immediately curved round to the

Joseph Locke, Engineer for the building of the Grand Junction Railway

north-west to pass to the north of Wolverhampton before running due north to Stafford. From Stafford the line resumed its north-western alignment to pass to the west of the Potteries. In general the route lay across favourable country and was designed with steeper gradients than the Liverpool and Manchester or the London and Birmingham;[5] consequently it required fewer major engineering structures. Within this chapter area, however, there are three notable viaducts: that at Aston, the long curved viaduct in the Rea valley which led to the Birmingham terminus, and a shorter viaduct at Penkridge.

The Grand Junction absorbed its progenitor, the Liverpool and Manchester, in 1845 and one year later merged with its contemporary, the London and Birmingham, and with the Manchester and Birmingham[6] (which ran only from Manchester to Crewe) to form the heart of the great London and North Western Railway.

In 1847 the Trent Valley line,[7] engineered by Robert Stephenson, George Bidder and Thomas Gooch, with Thomas Brassey as contractor, was opened to connect the Grand Junction Railway at Stafford with the London and Birmingham at Rugby, thus allowing traffic between London and the North West to avoid Birmingham and saving 9 miles.

Although the line was completed by June 1847, full services did not begin until December because of the necessity of reviewing the cast-iron structures on the line following the collapse of the Dee Bridge on the Chester and Holyhead Railway. Like the Grand Junction Railway, the Trent Valley line became part of the London and North Western Railway and it is still part of the West Coast Main Line.

1. WEBSTER N. W. *Britain's first trunk line, the Grand Junction Railway.* Adams and Dart, Bath, 1972.

2. ROSCOE T. *Book of the Grand Junction Railway.* Orr and Co., London, 1839.

3. RENNISON R. W. *Civil engineering heritage: Northern England.* Thomas Telford, London, 1996, 249–255 (HEW 223).

4. WHISHAW F. *The railways of Great Britain and Ireland.* Weale, London, 1842, 2nd edn, 124–137 (reprinted by David and Charles, Newton Abbot, no date).

5. LABRUM E. A. *Civil engineering heritage: Eastern and Central England.* Thomas Telford, London, 1994, 217–219. (HEW 1092).

6. Manchester and Birmingham Railway (HEW 1149).

7. Trent Valley Railway (HEW 1956).

R. CRAGG

Penkridge Viaduct, Grand Junction Railway

HEW 520
SJ 921 145

16. Penkridge Viaduct

Carrying the Grand Junction Railway over the River Penk and a minor road, Penkridge Viaduct was the first major railway work to be constructed by the eminent railway contractor, Thomas Brassey, and was completed in 1837.

There are seven arches of 30 ft span and the total length is about 240 ft. The piers and arches are constructed in brickwork with a stone cornice. There is a brick parapet wall. The piers are thickened as they near the ground, giving them a somewhat unusual curved profile.

17. Shugborough Tunnel

HEW 1126
SJ 981 216 to
SJ 988 216

As described in item 15, the Trent Valley Railway was constructed between Rugby and Stafford to avoid Coventry, Birmingham and Wolverhampton and reduce the direct mileage between Euston and Crewe. The Engineers for the line were Robert Stephenson and George Bidder, with T. L. Gooch as Chief Engineer.

Shugborough Tunnel, the largest engineering work on the Trent Valley line, is 777 yd long and carries a double line of railway under the flank of the Satnall Hills, through the grounds of Shugborough Hall. The tunnel has a semicircular arch springing from vertical walls, is brick-lined and built on a curve. The portals are built in ashlar masonry and have battlemented towers and, at the west portal, flanking walls leading to further ornate towers.

The contractor was Thomas Brassey in partnership with John Stephenson and William Mackenzie.

R. CRAGG

West Portal, Shugborough Viaduct

18. Holyhead Road

HEW 1213
SP 214 832 to
SJ 800 032

Watling Street, the Roman road from London to the Roman town of Viroconium (Wroxeter), near Shrewsbury, is followed closely by the present day A5 trunk road. A military road, it was beautifully constructed on direct alignments but with no regard for social and commercial interests.

The route chosen by Telford for his Holyhead Road across the area of this chapter left Watling Street at Weedon Bec (SP 632 599) and followed what is now the A45 road through Coventry to Birmingham; the A41 through West Bromwich, Bilston and Wolverhampton; then the A464 through Shifnal, to rejoin the line of Watling Street just west of Wellington. In 1919 the route was greatly improved for motor transport when the present A5 became the Holyhead Road and, running further to the north along Watling Street, avoided the Birmingham conurbation.

Telford's method of road construction[1] may be contrasted with those of the Romans as described in Chapter 5 and of John L. McAdam, described in Chapter 4. First he levelled and drained the line of the road, employing a maximum gradient of 1 in 30 wherever possible and laying cross drains every 100 yd along the road, connected with ditches on each side. On the prepared foundation was laid a pavement, 7 in. deep in the centre and 5 in. deep at the edges, of large stones placed by hand as closely as possible, with their broad ends downwards: the top stones were all no wider than 3 in. The interstices were hand packed with smaller stones. Then followed a further 6 in. of stones weighing not more than 6 oz and passing through a 2½ in. ring. The final surfacing was a 1½ in. thickness of gravel which soon, under the passage of wheels and horses' hooves, became a binding layer. A cross camber of about 4 in. was provided.

For lightly trafficked roads, Telford used the solid base pavement as before, but the tops of protruding stones were broken off to take a layer of stones broken to walnut size, followed by the gravel surfacing.

Telford's methods were more expensive, but resulted in a more durable road than those of McAdam. The Roman technique of incorporating mortar to strengthen the construction was of course available to Telford, but would have added even more to the cost. It must be remembered that the Roman roads were built by their troops or by conscript labour, whereas Telford had to employ paid workpeople.

1. TELFORD T. (ed. J. RICKMAN) *Life of Thomas Telford ... with a folio atlas of copper plates*. London, 1838, plate 82.

19. Staffordshire and Worcestershire Canal

HEW 1083
SJ 995 230 to SO 813 708

This is one of the original canals built in the latter half of the eighteenth century. The fact that it is still open to traffic serves to demonstrate that the early canals were built to satisfy a perceived need for water transport, in contrast to the later, speculative canals, many of which proved commercially unsuccessful.[1]

It is a narrow canal, designed for boats 70 ft long and 7 ft beam. It runs for 46 miles from the Trent and Mersey Canal at Great Haywood to the River Severn at Stourport, a town which owes its being to the canal. Begun in 1768 at its south end, with James Brindley as Engineer, it was completed in May 1772.

From just east of Stafford, which was reached in 1816 by a now derelict branch, the canal climbs southwards 100 ft by twelve locks to reach its summit level at Gailey Lock (SJ 920 106) where it passes under the A5 road. Just north of the end of the 10½ mile summit level at Compton Lock (SJ 884 989) are junctions with Telford's 1835 Birmingham and Liverpool Junction Canal at Autherley (SJ 902 020) and with the Birmingham Canal network at Aldersley (SJ 903 011). Until 1792 this afforded the only communication between Birmingham and the Severn. From the summit the canal begins a long descent by 29 locks to the valley of the River Stour, and follows the river closely to the Severn, 265 ft below the summit level.

At the junction between the rivers Stour and Severn Brindley built a series of basins and locks to connect his canal with the river, and around this sprang up the new town of Stourport-on-Severn (see Chapter 5).

1. LANGFORD J. I. *The Staffordshire and Worcestershire Canal*. Goose and Son, Cambridge, 1974.

20. Bratch Locks

HEW 982
SO 867 939

At Bratch, near Wombourne on the Staffordshire and Worcestershire Canal, is an interesting flight of three locks which between them lower the canal by 31 ft 2 in. in the course of its descent from the summit level into the valley of the River Stour.

The locks were designed by James Brindley and opened to traffic on 1 April 1771. Originally they were

Bratch Locks

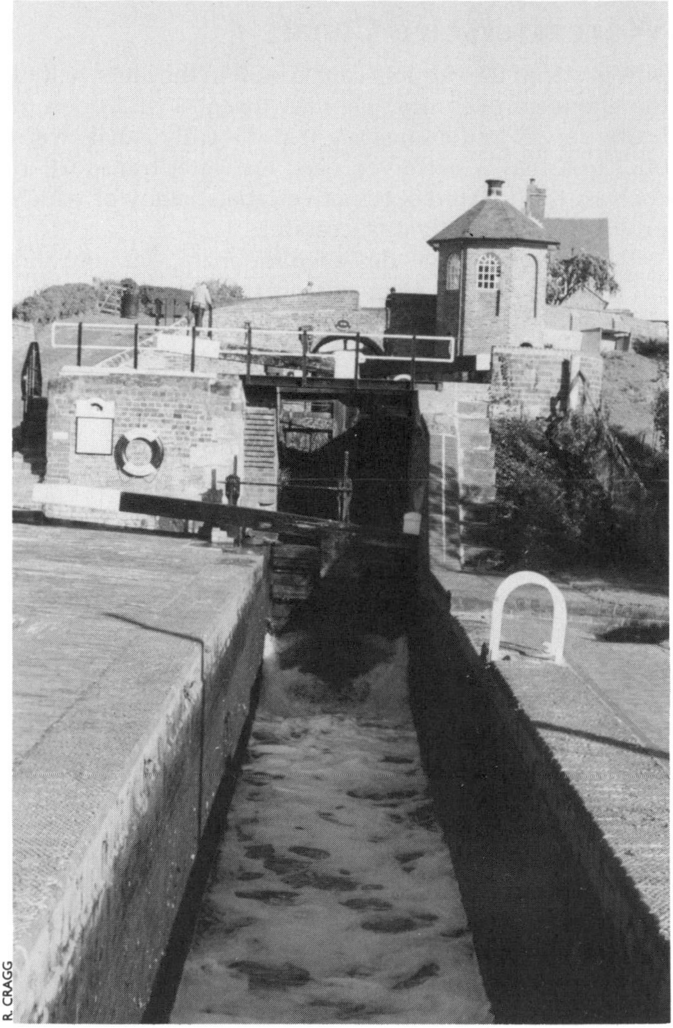

built as a staircase, with single deep gates separating the locks. Later they were converted to three separate locks by the addition of top gates to the two lowest locks. The length of the intermediate 'pounds' between the locks was thus extremely short, about 15 ft. Such a short length of canal could not accommodate the water discharged from the lock above without overflowing, so two long side ponds were constructed, connected to the canal by culverts through the canal walls. The side ponds are on the west side of the canal, the upper one extending for a considerable distance along the hillside.

The lock chambers, which are 70 ft long and 8 ft wide, reveal evidence of the alterations. The stonework of bridge 48 was cut to allow the new top gate of the middle lock to open fully and the chamber walls of the middle and lowest locks were recessed to receive the top gates when closed.

At the top of the locks is a neat octagonal toll-house, strategically sited to observe traffic approaching from both directions.

Under the arch of Bratch Bridge across the lower lock is an unusually complex set of access stairways in brickwork.

21. Trent Aqueduct, Great Haywood

HEW 885
SJ 994 229

Just south of the junction between the Staffordshire and Worcestershire Canal and the Trent and Mersey Canal at Great Haywood, this four-span stone aqueduct carries the Staffordshire and Worcestershire Canal over the River Trent. It was probably one of the later works to be built on the canal.

The technique of James Brindley, the Engineer of the canal, in building works of this kind was to construct one or more arches of the aqueduct on dry land, divert the flow of the river through them and then complete the aqueduct in the river bed. It is an excellent example of the massive structures necessary to carry the canal and its puddled clay bed which were built by the early canal engineers.

Between the aqueduct and the junction is a single arch over the site of the tail race from the nearby mill.

R. CRAGG

Aqueduct over the River Trent, Staffordshire and Worcestershire Canal

There is another similar aqueduct[1] carrying the canal over the River Sow not far away at Milford.

1. Milford Aqueduct (HEW 974) SJ 173 215.

22. Great Haywood Canal Junction Bridge

HEW 676
SJ 995 230

This elegant brick segmental-arch bridge carries the tow-path of the Trent and Mersey Canal over its junction with the Staffordshire and Worcestershire Canal with a very flat span of 35 ft 4 in., the rise of the arch being only 6 ft. The 20 in. thick arch ring is made up of two courses of brickwork with, over them, a 6 in. deep course of sand-stone blocks.

Designed by James Brindley, the bridge was built about 1772.

23. Trent and Mersey Canal

HEW 1135
SK 200 180 to
SJ 857 500

Among the earliest canals in Britain were those linking the four great rivers of central and southern England, the Mersey, Trent, Severn and Thames. One of the first of these was the 93 mile long Trent and Mersey Canal, then known as the Grand Trunk Canal and inspired by Josiah Wedgwood's concern for the transport difficulties then being experienced by himself and his fellow potters around Burslem and Stoke-on-Trent.[1]

The canal, from the Trent at Wilden Ferry (SK 459 309) near the mouth of the Derbyshire River Derwent, runs through the area covered by this chapter via Stone and Stoke-on-Trent; it rises 361 ft through 40 locks and passes through a short tunnel at Armitage (opened out in 1971), before passing under the high ground north-east of Stoke by way of Harecastle Tunnels. The north end of the canal terminates at a junction with the Bridgewater Canal at Preston Brook (SJ 568 810).

Work began in 1766 with James Brindley as Engineer, assisted by his Clerk of Works, Hugh Henshall, who took over on Brindley's untimely death in 1772. The canal was opened fully to traffic in May 1777 and continues in use largely unchanged to the present day, although commercial traffic is now almost entirely replaced by leisure craft.

Apart from short lengths at each end, the canal was designed for narrow boats, 70 ft long and 6 ft. 10 in. beam.

The canal connects with the national network at a

Great Haywood Canal Junction Bridge

R. CRAGG

number of places. At Great Haywood (SJ 995 230) it is joined by the Staffordshire and Worcestershire Canal and at Fradley (SK 140 140) near Lichfield a junction with the Coventry Canal[2] affords access to the Thames by way of the Oxford Canal.[3]

1. LINDSAY J. *The Trent & Mersey Canal*. David and Charles, Newton Abbot, 1979.

2. LABRUM E. A. *Civil engineering heritage: Eastern and Central England*. Thomas Telford, London, 1994, 242–243 (HEW 1090).

3. Ibid. 260–262 (HEW 1612).

24. Shugborough Hall Bridges

HEW 1787
SJ 993 227

In the grounds of Shugborough Hall, the home of Lord Lichfield, are two interesting iron bridges which carry estate paths over an arm of the River Trent.

The smaller bridge is the 'Chinese' bridge, so called because it is built adjacent to the Chinese House. It has a single cast-iron arch of 42 ft 6 in. span with two parabolic arch ribs. Each rib is 8 in. deep and 2 in. thick and the spandrel space is filled with rings of decreasing size towards the centre. The arch rib and spandrel rings are

'Chinese' Bridge, Shugborough Hall

R. CRAGG

all cast in one piece. The bridge is 6 ft wide and has a cast-iron parapet rail 2 ft 9 in. high with four ornate pillars surmounted by coronets at each end. The bridge was built in 1814.

The second, larger, bridge, built one year earlier in 1813, is a three-span cast-iron arched bridge with segmental arches of spans 19 ft 3 in., 20 ft 6 in. and 19 ft 3 in. The ribs, which are of P-section, are 6 in. deep, 2¾ in. thick at the top and 1½ in. thick at the bottom. The spandrel space is filled with cast-iron rings which are cast in one piece with the ribs. The two low stone piers are surmounted by cast-iron caps which carry cast-iron piers above. The cast-iron parapet rail is similar to the one over the Chinese bridge and is also terminated by decorative pillars.

HEW 1788
SJ 988 224

The ironwork for both bridges was supplied by John Toye and Company of Rugeley. No definite evidence of identity of the designer of the bridges has been located but Charles Heywood, who later became Clerk of Works on the Shugborough estate, was associated with their construction.

25. Essex Bridge, Great Haywood

HEW 1132
SJ 995 226

This ancient and historic sandstone masonry bridge, which spans the River Trent between Great Haywood and Shugborough Park, Staffordshire, has 14 arches of segmental form with voussoirs 14 in. deep. The total length of the bridge is about 310 ft, the spans of the arches varying between 14 and 15 ft. The line of the bridge is generally straight but at its south end there is a marked curve to the east, one of the arches being curved in plan to fit. The bridge is only 5 ft wide with stone parapet walls, giving a footway width of about 4 ft 3 in.

At each pier there are triangular cutwaters on both sides of the bridge and these are continued up to footway level to form pedestrian refuges 2 ft 3 in. deep and 4 ft wide.

The exact date of construction of the bridge is unknown but it is reputed to have been built at the end of the sixteenth century by the first Earl of Essex to carry his horses and hounds over the river on their way to Can-

R. CRAGG

Essex Bridge, Shugborough

nock Chase. It seems originally to have been much longer, since in the late seventeenth century there was a reference to a bridge having 43 arches compared with the present total of only 14. It is possible that the bridge continued as a causeway across the low-lying land bordering the river.

26. Wolseley Bridge

HEW 1786
SK 021 204

A crossing of the River Trent at the site of Wolseley Bridge, north-east of Rugeley, is recorded as early as the twelfth century. The present bridge, designed by John Rennie, dates from 1799 and is a three-span structure in sandstone ashlar which now carries the A51 road over the river. The bridge is built of stone obtained from the estate of Sir William Wolseley of Wolseley Hall, situated just to the east of the bridge, hence its name.

The arches are segmental with a centre span of 57 ft and two side spans of 52 ft 6 in. The arch ring is 36 in. thick. The substantial masonry piers are 9 ft 6 in. wide with cutwaters on both faces surmounted by decorative niches level with the springings of the arches. The river

R. CRAGG

Wolseley Bridge

flows through the southernmost two arches. The overall width of the bridge is 28 ft 6 in. and it carries a carriageway 20 ft 6 in. wide with small marginal haunches and no footways.

The construction of the bridge was started by James Trubshaw and Sons of nearby Colwich but the work was not completed. In April 1799 John Varley entered into a contract to complete the bridge but Rennie, who was supervising the work, considered it substandard and the work was finally completed by a Mr Potter.

27. Mavesyn Ridware Bridge

HEW 726
SK 092 168

The 'High Bridge', carrying the B5014 road over the River Trent between Mavesyn Ridware and Armitage, is one of two substantial cast-iron bridges built in east Staffordshire in the early nineteenth century under the direction of Joseph Potter, the County Surveyor.

The single cast-iron arch of 140 ft span was cast by the Coalbrookdale Company in 1830. It has five ribs, 36 in. deep and 2 in. thick, each rib having seven segments with bolted joints. The arch is given lateral stability by diagonal bracing running horizontally between the ribs and by transverse connections between the ribs at each segment joint. The spandrel is of X pattern made up from vertical and inclined members giving a 'latticed' appearance, which is enhanced by an arched member joining their intersections.

The roadway is 17 ft 9 in. wide and the total width of the bridge, including two narrow footpaths, is 25 ft 8 in. There was a cast-iron parapet rail 3 ft high. The stone abutments have wings which are curved in plan and incorporate massive stone pilasters.

The above description applied to the bridge as it stood from 1830 until 1982, when the bridge was threatened by mining subsidence. In that year a major operation was launched to safeguard the bridge during the expected period of subsidence. The road was diverted onto a Bailey bridge which was erected alongside and to the west of the bridge. Concrete piers were built in the river supporting a new steel arch which was positioned a few inches below the cast-iron arch. The weight of the iron arch was then transferred to the steel arch by jacks and packing. The old bridge was lightened by removing the road pavement and the handrail. In each rib a joint near the crown of the arch was unbolted so that relative movement of the original abutments could take place without fracturing the arch.

At the time of writing an operation is under way to construct a new bridge on the site of the Bailey bridge and to restore the old bridge for pedestrian use.

28. Chetwynd Bridge, Alrewas

HEW 168
SK 187 139

The second cast-iron bridge built under the direction of Joseph Potter carries the A513 road over the River Tame. It was built in 1824 and probably cast at Coalbrookdale.

There are three spans, of about 65 ft, 75 ft and 65 ft, the river passing though the two southernmost arches. Each arch has five ribs, each of five segments, with X-type spandrel bracing having a continuous arch-like member joining the intersections. There are transverse diaphragms at each segment joint. The roadway is 18 ft wide with a narrow footpath. There is a cast-iron handrail and massive stone abutments.

R. CRAGG

Chetwynd Bridge

In 1979 an inspection of the bridge revealed numerous fractures of the cast iron and an extensive repair operation was undertaken.

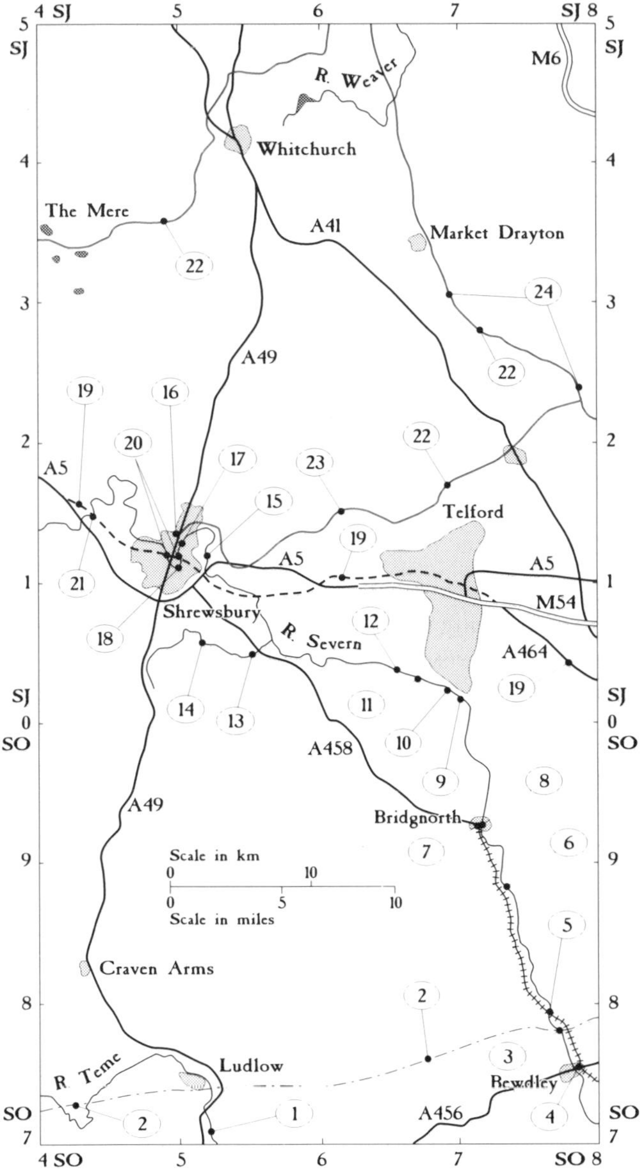

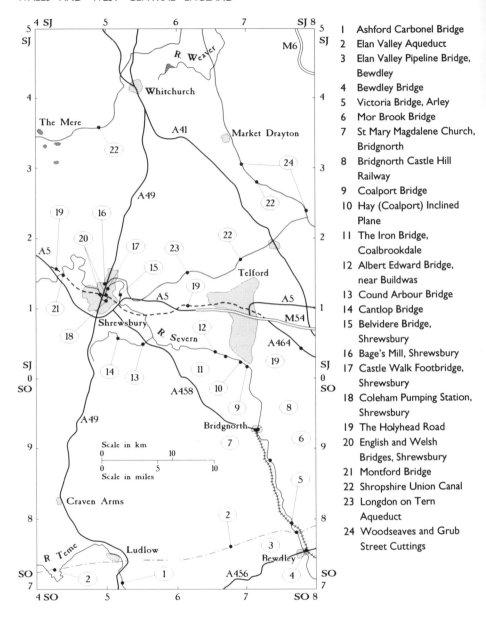

1 Ashford Carbonel Bridge
2 Elan Valley Aqueduct
3 Elan Valley Pipeline Bridge, Bewdley
4 Bewdley Bridge
5 Victoria Bridge, Arley
6 Mor Brook Bridge
7 St Mary Magdalene Church, Bridgnorth
8 Bridgnorth Castle Hill Railway
9 Coalport Bridge
10 Hay (Coalport) Inclined Plane
11 The Iron Bridge, Coalbrookdale
12 Albert Edward Bridge, near Buildwas
13 Cound Arbour Bridge
14 Cantlop Bridge
15 Belvidere Bridge, Shrewsbury
16 Bage's Mill, Shrewsbury
17 Castle Walk Footbridge, Shrewsbury
18 Coleham Pumping Station, Shrewsbury
19 The Holyhead Road
20 English and Welsh Bridges, Shrewsbury
21 Montford Bridge
22 Shropshire Union Canal
23 Longdon on Tern Aqueduct
24 Woodseaves and Grub Street Cuttings

7. Shropshire

Although now generally regarded as being of a rural character, Shropshire, for a relatively short time, was one of the cradles of the Industrial Revolution. Along the gorge of the upper Severn was found the necessary combination of raw materials (coal and iron ore), transport (the river) and inventiveness (Abraham Darby and his successors), which led to the development of the historic iron founding industry of the Coalbrookdale area.

The major influence of the Severn valley on industrial development is evident from a glance at the map, which shows the distribution of the civil engineering works described in the following pages. The river has acted as a major trade artery, attracting to its banks many industrial developments, but it has also formed a major barrier to communication from east to west, so that many of the works are concerned with crossings of the river by both road and rail. The provision of these river crossings has established a tradition of using the most modern technology of the day, starting with the first use of iron in bridge building at Ironbridge, and the early use of reinforced concrete in the Free Bridge at Jackfield (demolished in 1992) and culminating in the building of a new cable-stayed bridge to replace the latter.

Although many of the early civil engineers worked in the area, the name of Thomas Telford deserves particular mention. Early in his career, first as a budding architect and then as a major figure in civil engineering, he worked extensively in Shropshire and held several appointments there including that of Surveyor of Public Works for the county from 1787.

The more recent history of the region shows a marked reduction in its industrial importance. The coal and iron industry declined in the nineteenth century, while the neighbouring Black Country between Wolverhampton and Birmingham continued to prosper. The establishment of the new town, appropriately named after Telford, in the 1960s has helped not only to revitalize this area but also to encourage the preservation of many of the historic engineering works described in this book.

R. CRAGG

Ashford Carbonel Bridge

HEW 1101
SO 520 711

1. Ashford Carbonel Bridge

A fine single-span masonry segmental arch bridge carries the minor road from Ashford Carbonel to Ashford Bowdler over the River Teme south of Ludlow. Begun in 1795 and opened on 25 November 1797, the arch has a span of 81 ft and a rise of 24 ft 6 in. The voussoirs are mostly of gritstone with some limestone blocks and the spandrel walls are built in local red sandstone. The carriageway is 16 ft wide and there are no footways. A particular feature of the bridge is the paved invert between the abutments below river level.

The bridge was designed by Thomas Telford when he was County Surveyor for Shropshire and here he introduced for the first time the technique of hollow spandrel construction. This, by removing the need to fill the spandrel space with rubble, has the effect of reducing the total load on the arch and was used by Telford in several of his masonry bridges. The bridge was constructed by William Atkins the stonemason and Thomas Smith, carpenter, the contract being for £830.[1]

The bridge received major repairs in 1877 under the then County Surveyor, Thomas Groves, and was repaired again in 1970.

Just before 1795 building had started on an earlier bridge but this structure fell during construction, a mishap for which Adkins, the contractor, forfeited his bond.

1. BLACKWALL A. *Historic bridges of Shropshire.* Shropshire Libraries, Shrewsbury, 1985, 44–45.

2. Elan Valley Aqueduct

HEW 1194
SN 932 652 to
SP 003 804

The aqueduct carries water from Caban Coch reservoir in the Elan valley (see Chapter 2) a total distance of 73 miles to a storage reservoir at Frankley, south-west of the city of Birmingham.[1] Since this reservoir is about 170 ft lower than the inlet to the aqueduct at Caban Coch, the water is able to flow under gravity for the whole distance.

The general route is via Rhayader, Knighton, Ludlow, Cleobury Mortimer, Bewdley and Hagley to Frankley. For a total of one-half of its length the aqueduct acts as a conduit (that is the water flows at atmospheric pressure) and for the remainder, where the aqueduct is below the hydraulic gradient, the water flows in cast-iron or steel pipes under pressure.

In the initial scheme, first used in 1904 and officially opened in 1906, two 42 in. diameter pipes were installed and two more, each of 60 in. diameter, followed between 1919 and 1961.

In the construction of the aqueduct 15 tunnels were driven with a total length of 12.9 miles. The longest, Dolau Tunnel, is 4¼ miles long.[2] There are also several major structures which carry the aqueduct across roads, rivers, railways and other obstacles. Among the more notable are crossings of the River Teme at Leintwardine, Downton-on-the-Rock and Ludlow; the River Severn crossing at Bewdley (see below), and Hagley railway crossing.

The Engineer for the Elan Valley Aqueduct was James Mansergh.

1. MANSERGH E. L. and MANSERGH W. L. The works for the supply of water to the City of Birmingham from Mid-Wales. *Min. Proc. Instn Civ. Engrs*, 1911–12, **190**, 34–38.

2. LAPWORTH H. The construction of the Elan Aqueduct: Rhayader to Dolau. *Min. Proc. Instn Civ. Engrs*, 1899–1900, **140**, 235–248.

R. CRAGG

Elan Valley Aqueduct Bridge, Bewdley

3. Elan Valley Pipeline Bridge, Bewdley

HEW 1195
SO 775 782

The Elan Valley Aqueduct crosses the River Severn 2 miles north of Bewdley, on an impressive segmental steel arch of 150 ft span and 15 ft rise.[1] There are four arch ribs set at 12 ft 6 in. centres, each made up of riveted steelwork with top and bottom flanges separated by an X lattice. At the abutments each arch rib rests on an 8 in. diameter steel pin with a cast-iron shoe and heel plate bolted to a granite block.

There are three pipe galleries, originally designed to take two 42 in. diameter pipes in each. This is the lowest point on the route of the aqueduct and the water pressure is 250 lb/sq.in.

On the east bank of the river there is an approach viaduct over the flood plain with five segmental arches of spans varying from 34 ft to 62 ft. These are built in brickwork with stone facings. The total length of the viaduct and bridge is 624 ft.

1. MANSERGH E.L. and MANSERGH W.L. The works for the supply of water to the City of Birmingham from Mid-Wales. *Min. Proc. Instn Civ. Engrs*, 1911–12, **190**, 43–44.

4. Bewdley Bridge

HEW 461
SO 788 755

The present masonry road bridge at Bewdley is one of several which have crossed the Severn here since 1447. The first was destroyed in 1459 by the Lancastrians and was replaced in 1460 by a timber bridge which was rebuilt in 1483. This in turn was damaged in the Civil War, but was repaired in 1644 and remained until it was swept away by the great Severn flood of 1795, which did so much damage to bridges along the river.

The design for the new bridge was prepared by Thomas Telford. The summer and autumn of 1798 were very dry and the bridge was, astonishingly, built in that one period, the contractor being John Simpson of Shrewsbury. There are three main arches, the central arch having a span of 60 ft and a rise of 18 ft, and two side arches having spans of 52 ft and rises of 16 ft 9 in. The voussoirs have deep V-joints. There are two smaller arches on each bank of the river which carry the roadway over the river towpath and allow an easier passage for flood water. The overall width of the bridge is 27 ft and its balustraded stone parapet is 3 ft 10 in. high.

The road to the east of the bridge runs parallel to the river bank on a low viaduct of twelve semicircular arches surmounted by a neat cast-iron balustrade.

Bewdley Bridge

R. CRAGG

R. CRAGG

Victoria Bridge, Severn Valley Railway

HEW 464
SO 767 792

5. Victoria Bridge, Arley

The Severn Valley Railway, as its name suggests, follows the narrow valley of the River Severn and crosses it just south of the village of Arley. At this point was built the Victoria Bridge, with a single cast-iron arch of 200 ft span. It was opened for traffic on 31 January 1861 and carries a single line of railway.

The arch is made up of four ribs, each consisting of nine segments with bolted joints. An inscription cast into the centre of the arch records that it was designed by John (later Sir John) Fowler and cast by the Coalbrookdale Company. The contractors were Brassey, Peto and Betts of London.

The stone abutments of the bridge are pierced by brick arches which allow access along the banks of the river.

The line was closed to traffic in 1963 but the section of line from Bridgnorth to Bewdley and Kidderminster, including Victoria Bridge, has been reopened to traffic by a preservation company. Passengers may still cross the bridge in steam-hauled trains, a feature which has not been overlooked by television and film producers.

6. Mor Brook Bridge

HEW 972
SO 733 885

From ancient times the River Severn has been a trade route. In 1772 the towpath of the river from Bewdley to Coalbrookdale was, unusually, the subject of the creation of a turnpike trust. It was improved in the early nineteenth century to allow horses instead of manpower to be used for hauling boats.

Wherever a side stream entered the river, a bridge was necessary to carry the towpath over the obstruction. Mor Brook Bridge, about 4 miles south of Bridgnorth, on the west bank of the river, was built in 1824. It has a single cast-iron arch span with three ribs, each cast in two sections with a joint at the centre. The arch spans 30 ft between vertical brick abutments and the bridge is 5 ft 9 in. wide. Four-inch diameter tie rods are used to connect the arch ribs laterally. There is an attractive parapet of cast-iron X-lattice form with a decorative handrail.

The ironwork for the bridge was cast by John Onions of Broseley.

7. St Mary Magdalene Church, Bridgnorth

HEW 1134
SO 717 928

Thomas Telford, although best remembered as a civil engineer, had, in his early years, unrivalled experience as a master mason engaged on some of the finest monumental architecture of the time, such as Somerset House in London and Portsmouth Dockyard. Among several churches which he designed in the last years of the eighteenth century, one of the best examples is St Mary Magdalene, the parish church of Bridgnorth.

Prominently situated, the church has an unusual orientation, the nave lying north–south (rather than the traditional east–west). The tower is 120 ft high with a clock, eight bells and a copper-covered roof. The interior of the church is simple in design, the almost square nave being divided by two rows of Ionic columns with a timber gallery at the north end. The large arched windows in plain glass, extending the full height of the side walls, were intended to create (in Telford's own words) the appearance of 'one great and undivided apartment'.[1]

INSTITUTION OF CIVIL ENGINEERS

St Mary Magdalene Church, Bridgnorth

The church was built between 1792 and 1795 by John Rhodes and Michael Head.

Two other churches designed by Telford can be found in what is now the new town named after the great man. They are St Michael's[2] at Madeley, with its octagonal plan and square tower dating from 1796; and St Leonard's[3] at Malinslee, also octagonal, built in 1805.

1. TELFORD T. Letter to an unknown correspondent (undated). Telford Collection, Ironbridge Gorge Museum.

2. St Michael's Church, Madeley (HEW 1333) SJ 696 041.

3. St Leonard's Church, Malinslee (HEW 1780) SJ 689 081.

8. Bridgnorth Castle Hill Railway

HEW 1703
SO 717 930

Bridgnorth Castle Hill Railway is England's only inland cliff railway and runs between a low level station near the bridge over the River Severn to an upper station in Castle Walk.[1]

The total rise of the railway is 111 ft and the total length of the track is 201 ft, giving an average slope of 1 in 1.5 (66.7 per cent) which is claimed to be the steepest in Britain. To reach the upper station the line passes through a deep rock cutting with vertical sides. The track gauge is 3 ft 6 in., laid as a double line of rail with flat-bottomed rail on sleepers bolted to the underlying rock. There are twin cars, each with a capacity of 18 persons.

The railway was designed by G. Croydon Marks[2] and construction started in November 1891. The railway was first opened for passenger traffic on 7 July 1892.

R. CRAGG

Bridgnorth Cliff Railway

Originally the cars were connected together by two steel cables supported on rollers between the rails and the motive power was provided by water contained in a 2000 gallon tank beneath each car. The tank of the lower car was emptied and this provided sufficient out of balance force to operate the lift. The water was recirculated from the lower to the upper station by two gas engines supplied by J. B. Barker and Company of Birmingham. In 1944 the system was altered, each car being provided with a separate haulage rope connected to one of a pair of linked haulage drums together with a common safety rope running over the original head wheel. At the same time electrical haulage was introduced with a 32 h.p. motor at the upper station. The maximum speed of the cars is 250 ft per minute.

Tram-type coach-built cars were used, the car body being supported on a triangular framework of steel girders. In 1943 the cars were replaced, and replaced again by the present 'streamlined' bodies in 1955. Braking was originally by hydraulically operated brakes controlled by a brakesman on each car. These were subsequently replaced by brakes at the top station supplemented by automatic safety brakes on each car. The system now has air-operated brakes on the haulage system with emergency brakes on the cars.

1. GWILT C. F. *A history of the Castle Hill railway*. Rainbow Graphics and Print, Bridgnorth, no date.

2. MARKS G. C. Cliff Railways. *Min. Proc. Instn Civ. Engrs*, 1893–94, **116**, 318–326.

9. Coalport Bridge

HEW 422
SJ 702 021

The 103 ft span cast-iron arch bridge across the Severn, just below Coalport, was rebuilt into its present form by John Onions of Broseley in 1818 and is the third bridge on the site.[1,2] The first, built in 1780, had two timber arches with masonry abutments and a central pier, and is commemorated in the name of the adjacent Woodbridge Hotel. In 1799 the pier was removed and the timber arches replaced by a single span of three cast-iron arch ribs supporting a timber deck and parapets. The centre rib fractured in 1817.

IRONBRIDGE GORGE MUSEUM TRUST

Coalport Bridge

The present bridge includes some of the arch ribs from its predecessor, but the number of ribs was increased to five. It is one of the oldest cast-iron bridges still carrying traffic and is a scheduled ancient monument.

Coalport Bridge is a fine example of early cast-iron bridge building but its fame is rather overshadowed by its rather better known older cousin upstream at Ironbridge.

1. COSSONS N. and TRINDER B. *The Iron Bridge*. Ironbridge Gorge Museum Trust and Moonraker Press, Bradford-on-Avon, 1979, 79–86.

2. BLACKWALL A. *Historic bridges of Shropshire*. Shropshire Libraries, Shrewsbury, 1985, 19–22.

10. Hay (Coalport) Inclined Plane

HEW 639
SJ 694 026

The Hay Inclined Plane at Coalport, designed by William Reynolds, was built in 1792–93 to transport boats between the Shropshire Canal and the River Severn. The plane raised tub-boats in the dry on trolleys running on two parallel rail tracks, two boats being carried simultaneously, one on each track. As the main traffic of the canal was down to the river, the heavier boat descending drew up the lighter ascending boat, the trolleys being linked by ropes to a common winding drum on which was a brake wheel. A steam beam engine at the upper level was used solely to lift the loaded trolley over the sill of the upper basin, after which it descended under gravity.

The plane lifted the boats through a vertical height of 207 ft in a length of approximately 300 yd, a gradient of about 1 in 4. It could pass a pair of 5-ton tub-boats in 3½ minutes, instead of the 3½ hours which would have been spent in passing through the same height in conventional locks, possibly as many as 28 in number. The incline finally ceased to operate in 1907 and the rails were removed in 1910.

Shortly after the Ironbridge Gorge Museum Trust was formed in 1968, the upper basin and the incline were cleared. Restoration of the rail tracks and the lower basin followed in 1975–76.

11. The Iron Bridge, Coalbrookdale

'One of the boldest attempts with a new material was the application of cast iron to bridges.'

Thomas Tredgold, 1842[1]

HEW 136
SJ 672 034

The upper Severn gorge became one of the birthplaces of modern industry when in 1709 Abraham Darby, originally a Bristol brass founder, began to smelt local iron ore with coke made from local coal at Coalbrookdale. To enhance the communications of the area, a bridge over the River Severn was proposed between Broseley and Madeley Wood. Severe floods in earlier years suggested a single span to avoid piers in the river. In February 1776, a Bill was laid before Parliament for the construction of a bridge in cast iron. Thus was born the Iron Bridge, the first major bridge in the world to be constructed wholly of cast iron, and which gave its name, Ironbridge, to the settlement which sprang up around it.[2]

The bridge clears the river in a single arch of 100 ft span and is made up from ten half ribs, each cast in one piece by Abraham Darby III in his Coalbrookdale furnace. The project in essence is credited to the architect Thomas F. Pritchard. The detailed designs for the ribs and mem-

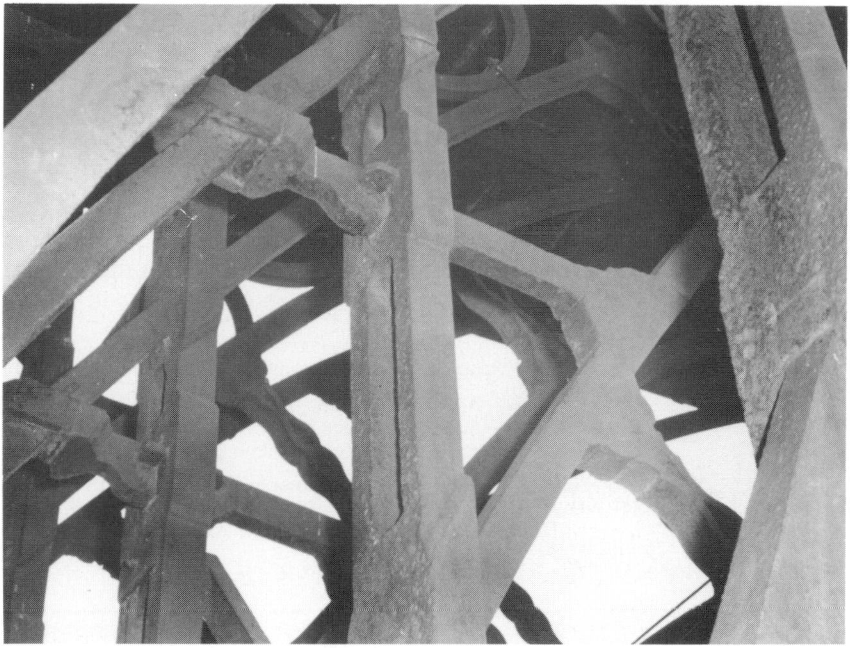

Joint detail, the Iron Bridge

bers of the bridge were prepared under the direction of Abraham Darby III, grandson of the first owner of the foundry, by Thomas Gregory, foreman pattern maker at Coalbrookdale and thus a worker in wood. This probably explains why the construction of the bridge is in the timber style, with mortise and tenon and dovetail joints.

Construction of the bridge took one and a half years and it was opened on New Year's Day 1781. It contains just over 378 tons of ironwork, probably equivalent to three or four months output from a contemporary furnace.[3] It survived flood and tempest to carry vehicular traffic until 1931 when it was closed to all but pedestrians. Major repairs were necessary at intervals during the life of the bridge, being occasioned mainly by the tendency of the sides of the gorge to move towards the river.

When, in 1796, Telford undertook a cast-iron bridge at Buildwas,[4] 2 miles upstream, he was aware of this tendency, since he states: 'I made the arch 130 ft span. The road rested on a very flat arch calculated to resist the abutments if disposed to slide inwards as at Coalbrookdale.' Alas, he was no more successful and the main arch broke in 1889, to be replaced by a steel bridge in 1905, which in turn was replaced by the present bridge in 1993.

In 1973, under the initiative of the Ironbridge Gorge Museum Trust, a reinforced concrete invert was constructed in the bed of the river between the two abutments of the Iron Bridge.[5]

The bridge and its toll-house on the south side are now preserved as a World Heritage Site and as a monument to the pioneering spirit of British engineers and craftsmen.

1. TREDGOLD T. *Practical essays on the strength of cast iron*. Weale, London, 1842, 4th edn, 10.

2. COSSONS N. and TRINDER B. *The Iron Bridge*. Ironbridge Gorge Museum Trust and Moonraker Press, Bradford-on-Avon, 1979, 11–36 and 48–52.

3. HODSON H. The Iron Bridge: its manufacture and construction. *Industrial Archaeology Review*, 1992, **XV**, 1, 36–44.

4. Buildwas Bridge (HEW 648) SJ 645 045.

5. COSSONS N. and TRINDER B. Op. cit. 119–125.

12. Albert Edward Bridge, near Buildwas

HEW 350
SJ 660 038

Opened on 1 November 1864, the Albert Edward Bridge carries the double line of the Wenlock Railway over the River Severn. It is very similar to the Victoria Bridge near Arley, with a span of 200 ft, and like it, was designed by John (later Sir John) Fowler and cast by the Coalbrookdale Company.

The four cast-iron arch ribs are in nine sections, with bolted joints, and spring from brick abutments. The decking is supported from the arch by cast-iron verticals, heavily braced. The original wrought iron and timber deck was replaced in 1933 by steel beams and plates, supporting ballasted track.

This is thought to be one of the last, if not the last, major cast-iron railway bridges to have been built and it is still in use today. Since the closure of the northern section of the Severn Valley line to which it formerly connected on the west bank of the river, the bridge now carries the branch line carrying the daily coal supply to the Ironbridge electricity generating station nearby.

R. CRAGG

13. Cound Arbour Bridge

Cound Arbour Bridge

HEW 423
SJ 555 053

The oldest cast-iron bridge in the county of Shropshire in use by modern traffic, Cound Arbour Bridge was built in 1797. It carries an unclassified road over the Cound Brook, 5½ miles south-east of Shrewsbury.[1]

The bridge spans 36 ft and comprises three cast-iron ribs manufactured at Coalbrookdale. The outer two are decorated by circles filling the space between the upper and lower edges of the ribs. Cast-iron plates carried directly on the ribs span the full 14 ft width of the bridge. The space between the plates and the road surface was originally filled with loose material, retained by cast-iron plates. In 1920 this filling was replaced by mass concrete and in 1931 new mass concrete abutments were provided. Since then the concrete, by acting as an arch, has helped the original ribs to carry the load of modern traffic.

1. BLACKWALL A. *Historic bridges of Shropshire.* Shropshire Libraries, Shrewsbury, 1985, 50.

R. CRAGG

Cantlop Bridge

HEW 330
SJ 517 063

14. Cantlop Bridge

Cantlop Bridge, also across the Cound Brook, was built in the year 1812 with a single span of 32 ft. It is the only survivor of four similar cast-iron bridges.[1] The bridge has four cast-iron arch ribs, each rib having a radial pattern of struts.

Although it is not known for certain who designed the bridge, it is closely associated with Thomas Telford, and it is now accepted as being the only surviving example of a cast-iron road bridge by him in the county. In his capacity as County Surveyor of Shropshire, he stated in a report to the Sessions of July 1812 that 'the bridge at Cantlop Ford although not enlarged as first suggested by [Telford] will according to the present plan be placed in such a footing as to be safely accepted as a County Bridge'. So well built was it that only small repairs and painting were needed over the next 163 years.

The ribs of the bridge are almost identical to, although shorter than, the ribs cast previously for similar bridges, all of 55 ft span: at Meol Brace, Shrewsbury (built in 1811 and dismantled in 1933); Cound (built 1818, dismantled 1969); and Stokesay (built 1823, dismantled circa 1965). It is suggested that the ribs for Cantlop Bridge were cast in a section of the same mould.

The bridge is scheduled as an ancient monument and in 1975 was bypassed by a new prestressed concrete bridge to ensure its preservation.

1. BLACKWALL A. *Historic bridges of Shropshire.* Shropshire Libraries, Shrewsbury, 1985, 49–50.

15. Belvidere Bridge, Shrewsbury

HEW 903
SJ 520 125

In 1849 this bridge was built to carry the twin tracks of the Shrewsbury and Birmingham Railway across the River Severn just east of Shrewsbury.

Each of its two skew spans of 101 ft 6 in. (89 ft on the square) is a cast-iron arch with six segmental ribs; the rise is only 10 ft 8 in. which gives an unusually high span to rise ratio of about 9.5 to 1. The spandrel spaces are filled with an X lattice and there are cast-iron floor plate units bolted to the top flanges and a cast-iron parapet. The abutments and 13 ft wide central pier are of masonry.

Apart from the renewal of deck plates, scour protection and grouting of the pier, the bridge required little maintenance from its construction until 1984, when a

Belvidere Bridge

R. CRAGG

reinforced concrete deck replaced the original cast-iron plates. The bridge continues to carry rail traffic with axle loads as high as 25 tons.

The arch sections were cast at Coalbrookdale, the Engineer was William Baker and the contractors were Hammond and Murray.

16. Bage's Mill, Shrewsbury

HEW 425
SJ 500 140

In 1796, work began on the construction of a flax mill in Castle Foregate on the northern outskirts of Shrewsbury. Powered by a 20 h.p. Boulton and Watt engine, the mill was completed in August 1797 to the design of Charles Bage, a Shrewsbury wine merchant, amateur engineer and friend of Telford, whose letters to William Strutt of Derby contain the earliest analysis of the strength of iron beams and columns. He had been brought into partnership, for his technical knowledge, by the Shrewsbury flax merchants Thomas and Benjamin Benyon, who were already in partnership with John Marshall of Leeds.

The mill is 174 ft long and 36 ft wide internally, and is five storeys high. The floors are carried on brick arches

Bage's Mill, Shrewsbury

IRONBRIDGE GORGE MUSEUM TRUST

springing from cast-iron beams, which are supported by the external brick walls and three rows of cruciform cast-iron columns. Hazledine's Shrewsbury foundry supplied the ironwork.

Bage's Mill was the world's first multi-storey building with an interior iron frame.[1] Its structure contained no combustible materials—an enormous advantage over the earlier type of textile mills with wooden floors. It was quickly followed by a succession of iron-framed mills at Salford, Belper, Leeds and elsewhere, with improved beam designs attributable largely to Bage himself.

In about 1897 the Shrewsbury mill was converted to a malthouse, but still retains its original structure. At the time of writing the building is no longer in use as a maltings but the present owners are discussing its conversion to other uses.

1. SKEMPTON A. W. and JOHNSON H. R. The first iron frames. *Architectural Review*, 1962, **131**, 175–186.

17. Castle Walk Footbridge, Shrewsbury

Castle Walk Footbridge,[1] one of numerous river crossings in Shrewsbury, is for pedestrians only; when opened in November 1951 it was the first prestressed concrete bridge to be built in Shropshire. It replaced a steel-wire rope suspension bridge, built in 1910, which had become unsafe because of corrosion. **HEW 1105** **SJ 499 130**

The bridge is formed from two balanced cantilevers, the ends of which support a central span, and is built of post-tensioned prestressed concrete. The main span is 150 ft and the two side spans are each 33 ft. The footway is 10 ft wide between the parapets.

The bridge was designed by architects T. P. Bennet and Company and consulting engineers L. G. Mouchel and Partners, in association with the Prestressed Concrete Company. The builders were Taylor Woodrow Construction.

1. BLACKWALL A. *Historic bridges of Shropshire*. Shropshire Libraries, Shrewsbury, 1985, 98–99.

IRONBRIDGE GORGE MUSEUM TRUST

Coleham Pumping Station

18. Coleham Pumping Station, Shrewsbury

HEW 424
SJ 496 121

Shrewsbury allowed all its drainage to run directly into the Severn until the 1890s, when a series of main collecting sewers was constructed parallel to the river banks to intercept and pick up the foul sewage. At Coleham, to the south-east of the town centre, the sewers on the north bank were connected to a cast-iron main sewer running underneath the river to a new pumping station on the south bank.

The brick-built station was equipped with twin beam pumping engines, each engine being a two-cylinder compound built by Messrs Renshaw. The two cylinders are vertical and are connected to one end of the overhead beam with a crankshaft and a 16 ft diameter flywheel at the other. Steam at a pressure of 90 lb/sq.in. was supplied from two boilers. The speed of the engine was 15 revolutions per minute, 114 gallons being pumped with each stroke.

The sewage was pumped through cast-iron pipes to a point in Whitehall Street from which it ran by gravity to the Sewage Treatment Works at Monkmoor. The whole system was opened on 1 January 1901.

The pumping engines continued in service until 1970, when electric pumps were installed, but the engines have been preserved and it is hoped to return one of them to steam working in the future.

19. Holyhead Road

HEW 1212, 1213
SJ 800 032 to SJ 400 179

Connecting the sections of the Holyhead Road already covered in Chapters 1 and 6, the route of the road through Shropshire commenced on what is now the A464 road near Albrighton, passed through Shifnal and rejoined the line of the Roman Watling Street east of Wellington, now part of Telford New Town. Westwards from Oakengates and Wellington the route followed, until recently, the modern A5 trunk road but improvements to the latter, such as the new alignment between Wellington and Shrewsbury and the bypass to the south of Shrewsbury, have departed from Telford's original route.

The Holyhead Road crossed the Severn in Shrewsbury on the English and Welsh bridges and then turned north-west to follow the Dee valley through Chirk and Llangollen and so on across Wales, as described in Chapter 1.

By 1829, Telford had completed his supervision of the rebuilding of this 267 mile long trunk road. Coaches could now travel over it at an average speed of 12 miles per hour—only a few years before the railways came to render such a speed a thing of the past.

20. English and Welsh Bridges, Shrewsbury

The historic town of Shrewsbury is built on a hill in the middle of a large loop of the River Severn, which from earliest times has necessitated river crossings both to the east and to the west.

The modern A5 trunk road, in bypassing the town to the south, has departed from the original route of the Holyhead Road, which now is the A458 through the town.

HEW 1026
SJ 496 124

The English Bridge is the eastern crossing. The first

SHROPSHIRE COUNTY LIBRARY

English Bridge, Shrewsbury

crossing here, in 1550, was a five-arch bridge, with a causeway of twelve culverts, which was replaced in 1774 by a new bridge designed by John Gwynn of Shrewsbury. This bridge, of Grinshill sandstone, was 400 ft long with seven semicircular arches, the central arch having a span of 55 ft. The overall width of the bridge was 23 ft 6 in. with a roadway only 15 ft wide.

In 1925 the complete rebuilding of Gwynn's bridge was begun under the direction of Arthur W. Ward, the Borough Surveyor.[1] The old structure was demolished and the stones were re-dressed and reused. The reconstructed bridge was given the same general configuration as its predecessor, with a central span of 54 ft 6 in., flanked on each side by spans of 47 ft 3 in., 41 ft 9 in. and 36 ft 6 in. in succession. By changing the central arch from semicircular to segmental form, the height at the crown of the centre arch was reduced by 5 ft and the overall width was also more than doubled to 50 ft between the parapets, with a reinforced concrete core and deck. The reconstructed bridge was formally opened by HM Queen Mary on 26 October 1927.

HEW 1025
SJ 489 127

The Welsh Bridge, on the west of the town, is the second on the site. The first was built about 1262 and was dismantled when the present bridge was opened, some

40 yd downstream, in 1794.[2] Traces of the old bridge are still visible on Frankwell Quay. The new bridge was designed and built by John Carline and John Tilley. Carline's father, also John, was foreman mason on the English Bridge. The overall length is 266 ft with five Grinshill sandstone arches. The span of the centre arch is 46 ft 2 in. and that of the others is 43 ft 4 in. The width between the parapets is 30 ft.

At mid-span, at the base of the downstream side parapet, there is a horizontal pulley wheel which was used for hauling boats over the shallows downstream of the bridge, the Severn at that time being navigable all the way up to Pool Quay, near Welshpool. The pulley wheel allowed the rope to be hauled at right angles to the river.

1. BLACKWALL A. *Historic bridges of Shropshire.* Shropshire Libraries, Shrewsbury, 1985, 97–98.

2. JERVOISE E. *The ancient bridges of Wales and Western England.* Architectural Press, 1936; republished by E.P. Publishing, East Ardsley, Wakefield, 1976, 132–135.

21. Montford Bridge

HEW 333
SJ 432 153

The first bridge to be designed by Thomas Telford after he was appointed County Surveyor of Shropshire, Montford Bridge formerly carried the A5 Holyhead Road over the River Severn 4 miles north-west of Shrewsbury. The bridge was built by John Carline (Junior) and John Tilley between 1790 and 1792, of local red sandstone quarried at Nesscliffe, and was opened on Lady Day 1792.[1]

The bridge has three segmental arches, the span of the central arch being 58 ft and the two outer spans 50 ft. The centre arch is 24 ft above mean water level. The original width of the roadway was 20 ft, but in 1963 the carriageway was widened to 22 ft and two footways, each 4 ft 6 in. wide, were provided. This widening was achieved by constructing a reinforced concrete slab, cantilevered 5 ft 6 in. beyond the existing width of the bridge on both sides. Approved by the Royal Fine Art Commission, this has not greatly detracted from the appearance of the bridge.

To superintend the construction of Montford Bridge, Telford sent for his old friend and workmate Mathew Davidson of Langholm; he also employed a mason from

R. CRAGG

Montford Bridge

Shrewsbury named John Simpson. This was the start of a lifelong association between the three men.

1. BLACKWALL A. *Historic bridges of Shropshire*. Shropshire Libraries, Shrewsbury, 1985, 24–26.

22. Shropshire Union Canal

HEW 1202–1206

The Shropshire Union Canal system resulted from the merger, circa 1846, of several canals and canal-owning railways.

It may be divided into five major groups, each with its various branches: the Ellesmere Canal, with the Chester Canal (covered in Chapter 8); the Birmingham and Liverpool Junction Canal (covered in Chapter 6); the Llangollen Branch of the Ellesmere Canal (covered in Chapter 1); the Montgomeryshire Canal (HEW 1205, SJ 255 203 to SO 116 921);[1] and the Shrewsbury Canal (HEW 1206, SJ 495 131 to SJ 707 125).[1]

In 1847 the London and North Western Railway Company leased the whole of the system. Since the early twentieth century there have been progressive closures of branches and sections of the main lines; but although much has disappeared, much also remains, even though

Thomas Telford

today it is only used for pleasure craft. Restoration is actively proceeding in various places.

The first of these canals to be opened was the Chester Canal to Nantwich in 1779. The main line of the Ellesmere Canal from the Chester Canal at Hurleston to Frankton was opened in 1805, as was the Llangollen Branch. The Montgomeryshire Canal opened in 1797 from Carreghofa in Shropshire (SJ 255 203), at the end of the Llanymynech Branch of the Ellesmere Canal, as far as Garthmyl (SO 194 944) south-west of Welshpool. Members of the Dadford family were its Engineers, but it was not until 1819, under Josias Jessop, that it reached Newtown on the Severn. A pair of cast-iron gates from the lock at Welshpool can be seen at the Waterways Museum at

Stoke Bruerne, Northamptonshire. The restoration of this canal is proceeding in stages.

The Shrewsbury Canal opened in 1796 over the 17 miles from Shrewsbury to Wellington and Oakengates (SJ 672 126). Josiah Clowes was succeeded as Engineer by Thomas Telford in 1795. The more important works on this canal included the iron aqueduct at Longdon on Tern; an inclined plane with a 75 ft lift at Trench; and the 970 yd long Berwick Tunnel, the first tunnel to have a towpath, which took the form of a 3 ft wide timber gangway on bearers set in the wall. This canal is now completely derelict but luckily the Longdon on Tern Aqueduct has survived.

At Wellington, the Shrewsbury Canal joined the 1788 Shropshire Canal which, with three inclined planes, led to the Severn at Coalport, some 22 miles below Shrewsbury. There were also connections to a network of small canals in the area[2] including the Donnington Wood Canal (1768), the Wombridge Canal (1788) and the Ketley Canal (1788). All these canals have been disused for many years and have now almost completely disappeared under the New Town of Telford; however, the inclined plane at Coalport, together with a short section of the Shropshire Canal, has survived.

The last of the component canals to be opened was the Birmingham and Liverpool Junction Canal which was finally completed in 1835, linking the Birmingham canal system at Autherley Junction to a junction with the Chester Canal near Nantwich. Its Newport branch connected with the Shrewsbury Canal at Wappenshall Junction and so joined the network of Shropshire canals, which until then had been isolated, to the national canal system.

1. Morriss R.K. *Canals of Shropshire*. Shropshire Books, Shrewsbury, 1991.

2. Hadfield C. *The canals of the West Midlands*. David and Charles, Newton Abbot, 1985, 150–165.

23. Longdon on Tern Aqueduct

HEW 280
SJ 617 156

The Longdon on Tern Aqueduct, designed by Thomas Telford, carried the Shrewsbury Canal over the River Tern, a small tributary of the Severn, close to the B5063

Longdon on Tern Aqueduct

Wellington to Shawbury road, about 5 miles north-west of Wellington.

Built in 1796, it was the first canal aqueduct to be designed in cast iron, and has a trough 9 ft wide and 3 ft deep, and is 186 ft long with four spans of 47 ft 8 in. The towpath is carried alongside the south side of the trough, level with its base. The slender lines of the trough, and its three-legged vertical and inclined cruciform cast-iron supports, are in strange contrast with the heavy masonry abutments, each with two flood arches. The reason for this may be found in the Shrewsbury Canal Company's minute book which records that Mr Clowes, the Engineer, had died in 1795 and that Thomas Telford had been appointed to succeed him. It would seem that Clowes had already completed a masonry aqueduct which was either unsatisfactory to Telford or had been damaged by floods, and this new form of iron structure replaced it.

The canal was closed in 1944 but the aqueduct remains in its original location, although the canal embankment which served it has been levelled.

24. Woodseaves and Grub Street Cuttings

HEW 2070
SJ 692 323 to SJ 701 299

In the design of the Birmingham and Liverpool Junction Canal, its Engineer, Thomas Telford, employed techniques which had more in common with railway construction than with canals. As described in Chapter 6, the canal pursues a bold straight path across the country and this required the provision of large-scale earthworks, deep cuttings and high embankments.

The two cuttings described here are typical of those found along this canal.

Woodseaves Cutting is about 2900 yd long (1.65 miles) and up to about 70 ft deep with steep sides, almost vertical in places. In the deepest part of the cutting the waterway narrows to about one half its normal width for a distance of 1800 yd; this was presumably done to economize on excavation. The cutting is crossed by four bridges. Two of these are of the type normally encountered on canals but the other two, the appropriately named High Bridge (No. 57 at SJ 697 307) and Hollings Bridge (No. 58 at SJ 693 316), are of an unusual design, being formed from relatively small-span semicircular arches situated at the top of very high vertical abutment walls. In the case of Hollings Bridge the top of the bridge is 40 ft above the water level and High Bridge is even higher.

HEW 2062
SJ 778 255 to SJ 792 234

The rather unattractively named Grub Street Cutting, situated about 7 miles further south, is actually just over the border in Staffordshire but is included here for the sake of completeness. It is about the same length, 2900 yd, but is of slightly less depth and has shallower side slopes. The maximum depth is about 50 ft and it is crossed by three bridges, of which two show the characteristic high abutment walls also encountered in Woodseaves Cutting. High Bridge (again!) (No. 39 at SJ 790 242) is just over 40 ft above the waterway and Double Culvert Bridge (No. 40 at SJ 790 246) is 32 ft high. In the case of High Bridge, which carries the A519 road, there is another unusual feature in that at some time after the building of the bridge the high abutment walls have been buttressed by the addition of a double brick arch structure about halfway up the walls. At one time this canal carried along its bank a line of telegraph poles and wires and the strengthening arch has been used as a convenient siting point for a very short telegraph pole. Although the other poles have now been removed (and traces of them may still be found along the towpath), this one has been left as a curiosity.

Another deep, steep-sided cutting together with a short tunnel can be found at Cowley, a further 5 miles south.

The digging of these cuttings presented many problems to the contractor, William Provis, because of the frequent slipping of the cutting sides which, with present-day knowledge of soil mechanics, should have been made much shallower, albeit at the cost of much additional excavation and land. Even today, slips continue to take place from time to time and evidence of this can clearly be seen from a towpath walk through the cuttings.

Both cuttings were completed by about 1832 but the final opening of the canal was delayed by the difficulties encountered in the building of the great embankment at Shelmore, described in Chapter 6. Boats were not able to use the entire length of the canal until March 1835.

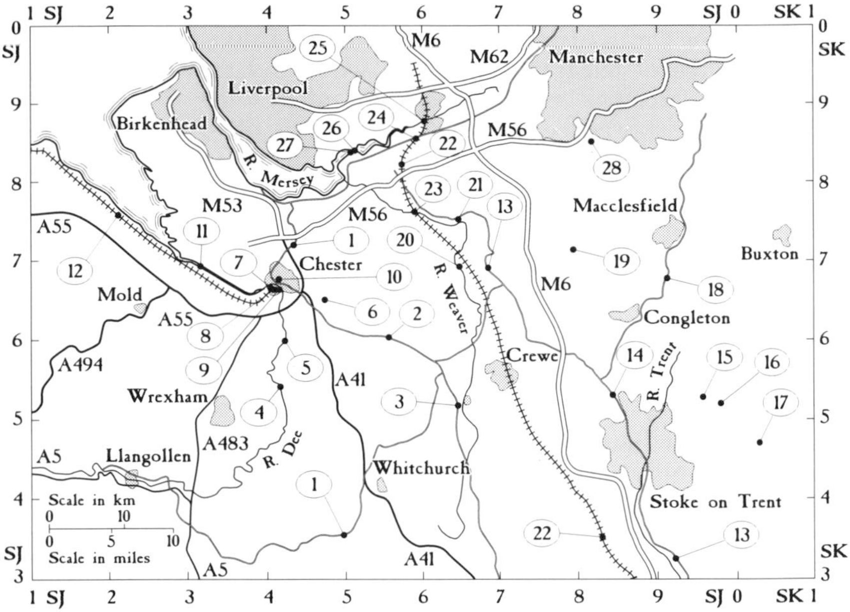

1 The Chester and Ellesmere Canals
2 Beeston Iron Lock
3 Nantwich Cast Iron Aqueduct
4 Holt Ancient Bridge, Farndon
5 Eaton Hall Iron Bridge
6 Hockenhull Packhorse Bridge
7 Chester Weir
8 Old Dee Bridge, Chester
9 Grosvenor Bridge, Chester
10 Chester Water Tower
11 River Dee Channel
12 Chester and Holyhead Railway
13 Trent and Mersey Canal
14 Harecastle Tunnels
15 Hazlehurst (Denford) Aqueduct
16 Cheddleton Mills
17 Froghall Basin
18 The Macclesfield Canal
19 Jodrell Bank Radio Telescope
20 The Weaver Navigation
21 Anderton Boat Lift
22 Grand Junction Railway
23 Dutton Viaduct
24 Acton Grange Viaduct
25 Warrington Bridge
26 Runcorn–Widnes Road Bridge
27 Runcorn Railway Bridge
28 Manchester Airport

8. Cheshire and North Staffordshire

Cheshire is basically a shallow plain resembling a dish, which is split near the middle by a range of hills. The county is bounded on its eastern flank by the Pennines, and on its western flank by the Clwydian Range; the median sandstone ridge is crowned by Delamere Forest. The northern geographical boundary is the River Mersey.

Cheshire has been an important corridor for trade since prehistoric times, and was of great military importance because of its proximity to North Wales and the routes which led north to the Scottish border. Chester now stands where the Roman legionary fortress of *Deva* was built, and in medieval times the bridge over the River Dee was maintained by the Crown because of its strategic importance.

The industrial development of Cheshire and the county's salt and allied chemical industry was considerably aided by the construction of the canal network, giving access to sea routes and thus to new markets. From the eighteenth century the development of the turnpike road system, in which such eminent figures as Telford, Metcalfe and McAdam were involved, further assisted this process and during the nineteenth century the spreading railway network again promoted industrial growth. Crewe grew into one of the largest railway towns in England, and Chester itself was the hub of six railway routes. In recent years motorway construction has further improved communications for the industrial areas and provided easier access to the North Wales scenery for the development of tourism.

In the south-east of the region covered by this chapter lies the part of north Staffordshire dominated by the intensively industrialized area centred on Stoke-on-Trent and bounded to the east by the Pennine ridge. The county boundary between Cheshire and Staffordshire lies squarely on the watershed, the Cheshire rivers flowing north to the Weaver and the Mersey, while the River Trent flows south through the Potteries. Within a relatively narrow corridor are found a variety of transport links, including the early Grand Trunk Canal (now the Trent and Mersey) of 1777, with its great tunnels at Harecastle, the Grand Junction Railway, built 60 years later, and, more recently, the M6 motorway.

1. Chester and Ellesmere Canals

HEW 1202
SJ 370 318 to
SJ 405 774

These canals eventually formed part of the Shropshire Union Canal system (see Chapter 7). The 19¼ miles of the Chester Canal from the Dee at Chester to Nantwich were built under the direction of Samuel Weston for boats 72 ft long and 13 ft 6 in. wide, and were opened in 1779. The intention was to link with the Trent and Mersey Canal at Middlewich, but this was not achieved until 1833 when a branch from Wardle (SJ 613 571) opened. The Chester Canal was therefore for some years in financial straits.

The Ellesmere Canal was authorized under an Act of 1793. Thomas Telford, then aged 36 and County Surveyor of Shropshire was, to use his own words, 'appointed Sole Agent, Architect and Engineer to the Canal',[1] with William Jessop as Consulting Engineer. It was proposed that the canal would link the Mersey, Dee and Severn rivers via Ellesmere Port and Chester, passing through the North Wales coalfield and by way of Wrexham, Ruabon, Chirk and Frankton (near Ellesmere) to Shrewsbury.

Work began on the 8¾ mile stretch from Chester to Netherpool (later Ellesmere Port), known as the Wirral Line: this opened in 1797 and merged with the Chester Canal in 1813 to become the Ellesmere and Chester Canal. Work began at the same time on part of the main line which eventually became the Llangollen Branch.

In 1800 a less expensive route from Chester was adopted, from the Chester Canal at Hurleston (SJ 626 553) north of Nantwich to Frankton (SJ 370 318), passing to the west of Whitchurch; the section of canal north of Ruabon was never built. However, the originally planned north–south route required the building of the magnificent Pontcysyllte Aqueduct, on which construction was already well advanced when the decision was taken to change the route (see also the Shropshire Union Canal, Llangollen Branch in Chapter 1). Fortunately for posterity, that marvellous structure was completed. In addition, the Ellesmere Canal Company also failed to complete its intended line to the Severn near Shrewsbury, the canal south-east of Frankton petering out in the fields near Weston Lullingfields.

The canal port complex at Ellesmere Port,[2] now leased

to the borough council, has been rehabilitated and now forms the Ellesmere Port Boat Museum.

The Northgate Locks[3] in the city of Chester are of interest because they are cut out of solid rock.

1. TELFORD T. Letter to Andrew Little of Langholm from Shrewsbury, 29 Sept. 1793. Copy in Telford Collection, Ironbridge Gorge Museum.

2. Ellesmere Port Canal Terminus (HEW 1287) SJ 405 772.

3. Northgate Locks, Chester (HEW 1337) SJ 401 666.

2. Beeston Iron Lock

HEW 456
SJ 554 599

The Ellesmere and Chester Canals had, from the start of their construction, suffered from local failures caused by pockets of silt and running sand, and in 1787 Beeston Lock collapsed because of these unstable conditions. Traffic ceased for a time because there were no funds available for repair.

In 1827 a short section of the canal was realigned and two new locks were built, the lower one adjacent to the road (now A49) being constructed entirely of flanged iron plates bolted together, as in the Pontcysyllte Aqueduct. This unique method of construction, devised by Thomas

Beeston Iron Lock

R. CRAGG

Telford, proved satisfactory and no further stability problems were encountered.

3. Nantwich Cast Iron Aqueduct

HEW 497
SJ 642 526

Telford's aqueduct, constructed about 1830, carries the Ellesmere Canal over the Nantwich to Chester road with a headroom of 15 ft 6 in. and a clear span of 29 ft 6 in. between the masonry abutments. Each side of the aqueduct proper is constructed of 6 ft square flanged cast-iron plates bolted together, providing a waterway of 13 ft with a 4 ft brick-surfaced towpath. The whole is surmounted on both sides by cast-iron railings 3 ft 9 in. high.

With the better known aqueduct at Stretton over the A5 road and a third at Congleton[1] on the Macclesfield Canal, these works form a trio all designed by Telford to the same general pattern.

1. Congleton Aqueduct (HEW 494) SJ 866 622.

Nantwich Aqueduct

R. CRAGG

R. CRAGG

4. Holt Ancient Bridge, Farndon

Holt Ancient Bridge

HEW 1221
SJ 412 544

This bridge crosses the River Dee on the border between England and Wales and between the villages of Farndon and Holt. It has eight arches of about 24 ft span and a total length of 520 ft, which includes the lengthy approach on the Welsh side. The width between parapets is only 13 ft, of which the carriageway takes up 10 ft. Seven of the arches are of normal two-ring construction, but the sixth arch from the English side has a further arch ring about 3 ft above the voussoirs. This suggests that there was a defensive tower over this span, which is known locally as 'The Lady's Arch', and also that the tower included a shrine dedicated to Our Lady. The date of the bridge is variously reported as 1345 and 1545.

5. Eaton Hall Iron Bridge

HEW 856
SJ 417 601

The private drive to Eaton Hall is carried across the River Dee on a cast-iron bridge constructed in 1824 a few yards upstream from the old ford (Aldford), by which the Roman Watling Street crossed the river. The single span of 151 ft consists of four arched ribs of I section, 36 in.

P. DUNKERLEY

Eaton Hall Iron Bridge

deep in seven segments, with ornate cruciform lattice bracing in the spandrels.

The bridge carries the names of several of those who were involved in its design and construction. They are William Crosley ('Surveyor'), William Stuttle ('Clerk of Works'), William Stuttle Junior ('Founder') and William Hazledine ('Contractor'). Hazledine, the ironmaster, worked with Thomas Telford on Pontcysyllte Aqueduct and many other bridges. The excellent quality of the bridge is no doubt attributable to its aristocratic connections.

The 36 in. wide cast-iron deck plates extend the full 17 ft width of the bridge and support some 14 in. of road metal.

The site is also an important point for gauging the flow of the River Dee.

6. Hockenhull Packhorse Bridge

HEW 939
SJ 476 657

There are two packhorse bridges to the south-west of Chester—near Wrexham at Caergwrle and at Ffrith. However, perhaps the bridge 4 miles east of Chester at Hockenhull is the more worthy of mention here, as it lies

R. CRAGG

Hockenhull Packhorse Bridge

on the old route from Chester to London which ran rather to the south of the present A51 road. The River Gowy is crossed at Hockenhull by three separate arches, spanning 12 ft to 17 ft in a total length of construction of some 240 ft.

It is shown on John Ogilvie's map of 1675 and remained in use until 1769 when it was bypassed. It is thought to be of seventeenth-century construction, although there is reason to believe that there was a bridge at the site before 1353.

7. Chester Weir

HEW 604
SJ 407 658

In 1071 a Norman, Hugh d'Avranches (Lord Lupus), became Earl of Chester and it was he who built the first weir, on the eastern side of the present Old Dee Bridge, to provide the head of water necessary to power the Dee mills. During 1281 the mills were destroyed and the weir was damaged by flood, but they were quickly rebuilt by Richard the Engineer, Master Mason at Chester, as they belonged to the King and were the most important of the many mills in the county.[1] They even became famous nationally and are remembered in the song 'The Jolly Miller of the Dee'.

The Dee mills were destroyed by fire in 1895 but the height of the weir was raised early this century and used to power the hydroelectric generating plant which was constructed on the site.

This plant no longer operates, but the weir, which is approximately 170 yd long, still remains to give security to navigation upstream. By preventing tidal waters from flowing above the weir, it also protects water supplies from salt. It was, at one time, the principal flow gauging point on the Dee.

1. TURNER R. Notes on the life and career of Richard the Engineer. Unpublished typescript.

Chester Weir

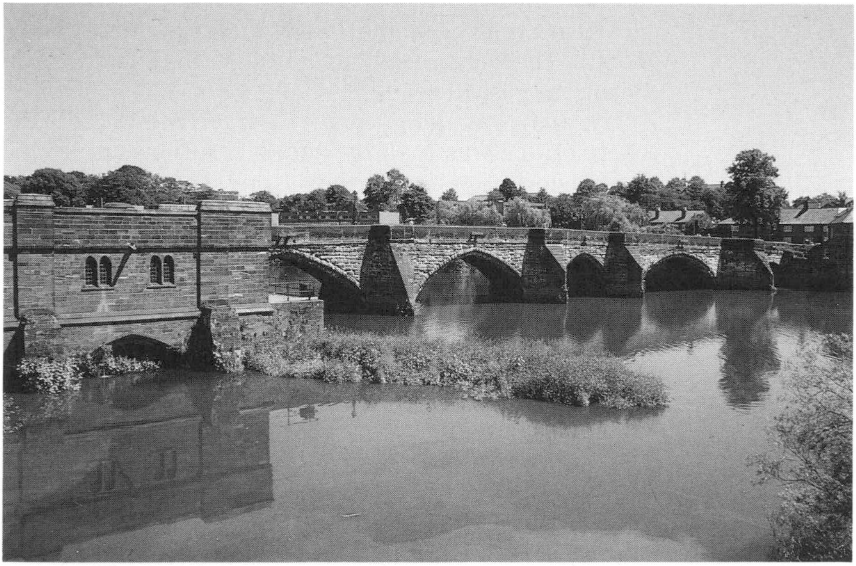

R. CRAGG

Old Dee Bridge, Chester

8. Old Dee Bridge, Chester

HEW 105
SJ 407 657

This red sandstone bridge has seven arched spans varying from 23 ft to 60 ft, one being semicircular, two segmental and four pointed. They surmount broad cutwaters, most with recesses above. The present structure was built during the fourteenth century by Henry de Snelleston, mason and surveyor to the Black Prince. It has been rebuilt or repaired on many occasions since then. The last major rebuilding was in 1826, when a footway was added on its eastern side. Until 1832 it was the only bridge across the River Dee at Chester.

9. Grosvenor Bridge, Chester

HEW 104
SJ 402 656

Thomas Harrison was the architect for Grosvenor Bridge about the year 1802, but its construction was delayed for 25 years, by which time he was nearly 80 years old. It is the crowning glory of his career:[1] when it was formally opened by Princess, later Queen, Victoria in 1834, it was the longest stone arch in the world, with a clear span of 200 ft.[2,3] It is still the longest in Britain and is surpassed in Europe by only three others, all built in the twentieth century: one at Plauen (295 ft) in Germany; one at Salcano

(279 ft) in Croatia; and the Pont Adolphe at Luxembourg (276 ft).[4] There are seven longer bridges in China, all built during the past 40 years.[5]

The arch was designed by George Rennie[6] and the whole bridge was built by James Trubshaw under the supervision of Jesse Hartley, using Peckforton stone, Scottish granite and some Chester red stone. The bridge has a rise of 42 ft and carries a 24 ft wide carriageway between the parapets, which are 33 ft apart.

The best view of the bridge is obtained from Edgar's Field on the south bank of the river, whence the Old Dee Bridge can also be seen.

1. BACHE A. Correspondence: The aesthetic treatment of bridge structures, by A. J. Husband. *Min. Proc. Instn Civ. Engrs*, 1900–01, **145**, 217.

2. HARTLEY J. An account of the new or Grosvenor Bridge over the River Dee at Chester. *Trans. Instn Civ. Engrs*, 1836, **1**, 207–214.

3. *British Bridges*. Public Works, Roads and Transport Congress, London, 1933, 33–34.

4. SÉJOURNÉ P. *Grands voûtes*. Tardy-Pigelet, Bourges, 1913–16: **2**, 67–82; **3**, 29–31, 52–58 and 141–149.

5. XIANG HAIFAN (ed.) *Bridges in China*. Tongji University Press, Shanghai, 1994.

6. Institution of Civil Engineers. Memoir on G. Rennie. *Proc. Instn Civ. Engrs*, 1868–69, **28**, 611–612.

Grosvenor Bridge, Chester

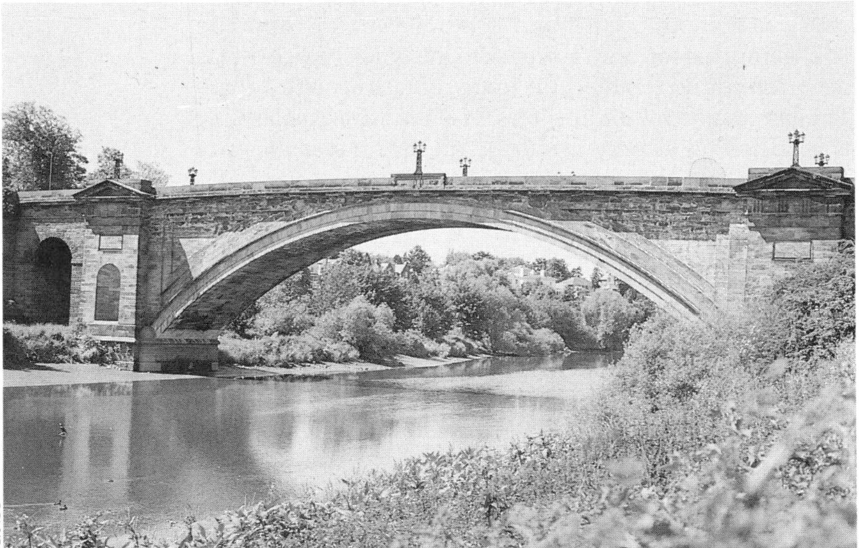

R. CRAGG

R. CRAGG

Chester Water Tower

10. Chester Water Tower

HEW 1080
SJ 416 666

Since the incorporation of the City of Chester Waterworks in 1826, water has been taken from the River Dee at the Barrel Well site, and until 1853 it was pumped untreated into the supply system.

The present Tarvin Road Waterworks was then built to the designs of J. F. La Trobe Bateman. The works included a reservoir, three slow sand filters, a pure water tank, a pumphouse and a water tower. The last two and one of the sand filters still exist.

The 70 ft diameter brick tower was originally 64 ft high, surmounted by a 12 ft deep cast-iron tank with a capacity of 268 000 gallons. In 1889, a further 20 ft head of water to meet the demands of the expanding city was achieved by jacking up the tank and inserting additional brickwork between the original top of the tower and the base of the tank.

11. River Dee Channel

HEW 2047
SJ 396 666 to
SJ 264 732

In AD 50 the Romans created the port of Deva (Chester) and made improvements to the River Dee as far up as Farndon (SJ 412 544). Chester was the principal port of the north-west of England until the fifteenth century, but siltation and the need to accommodate larger vessels began to have an adverse effect. The main shipping activities then took place at newly created wharves at New Haven (Little Neston) and later at Parkgate (SJ 280 778)[1] where some visible evidence still exists.

Various attempts were made over the years to improve the Dee from the estuary to Chester.[2,3] The successful proposal by Nathaniel Kinderley was authorized by an Act of 1732. This involved the construction of a new channel with a depth of 16 ft at moderate spring tides, to allow the passage of 200 ton vessels. The channel was 80 ft wide and the total length, from the Water Tower in Chester to Connah's Quay, was 14 000 yd (8 miles).

The new channel obstructed the existing fords across the Dee and ferries were instituted at Saltney and Queensferry. Chester was successful again as a port for possibly 150 years but had virtually closed by the early 1900s. The ferries were replaced by bridges—at Queensferry by a road bridge, the first being in 1897, and at Saltney by a footbridge in 1969. The channel still exists, is tidal and is navigable by small craft.

Following the excavation of the new channel extensive areas of land were reclaimed from the estuary at various times from 1740 until 1916. During this period the River Dee Company was advised by many of the eminent civil engineers of the day including Thomas Telford, Sir John

Rennie and Henry Robertson, but not all the work proposed was carried out.

A very full account of the reclamation work on the Dee estuary is given by Webster.[4]

1. PLACE G. *The rise and fall of Parkgate's passenger port for Ireland 1668–1815.* Carnegie, Preston, 1994.

2. WILLAN T. S. *River navigation in England.* The Chetham Society, 1964, 64–67.

3. WRIGHT G. A. Canalization of the River Dee. *Proceedings of the Flintshire Historical Society,* 1967, Nov.

4. WEBSTER F. The River Dee reclamations and the effect upon navigation. *Transactions of the Liverpool Engineering Society,* 1930, **60**, 63–92.

12. Chester and Holyhead Railway

HEW 1094
SJ 413 760 to
SH 248 822

The Chester and Holyhead Railway, already dealt with in some detail in Chapter 1, began at Chester General Station. The imposing range of station buildings by Francis Thompson, although altered several times, still remains. The contractor for the first 8 miles of the railway from Chester was Edward Betts.

After leaving the station, the line passed through a tunnel which was later opened out to form a cutting, and two more tunnels, that under Windmill Hill being unusual in that it accommodates four tracks. The Chester and Holyhead Railway then crosses the River Dee on a three-span girder bridge, originally cast iron with wrought iron bracing. On 24 May 1847 one of the 98 ft spans collapsed under a train. The inquiry which followed did much to hasten the development of straightforward wrought iron and, later, steel plate girders.

13. Trent and Mersey Canal

HEW 1135
SJ 857 500 to
SJ 567 810

The Trent and Mersey Canal, which has been dealt with more extensively in Chapter 6, passes under the high ground between Kidsgrove and Tunstall, north-east of Stoke-on-Trent, by means of the Harecastle Tunnels, before beginning the descent via Middlewich and Northwich with 36 locks (originally 35) to the Bridgewater Canal[1] at Preston Brook (SJ 567 748).

There are also tunnels at Barnton, 572 yd long

James Brindley

(SJ 634 748), Saltersford, 424 yd long (SJ 626 753) and Preston Brook, 1239 yd long (SJ 572 794).

The Caldon Canal,[2] which runs from the Trent and Mersey Canal at Stoke-on-Trent to Froghall (SK 022 475), was opened in 1778. A junction is also made just north of Harecastle Tunnels with the Macclesfield Canal which provides a link to Manchester via the Peak Forest Canal at Marple.

1. RENNISON R. W. *Civil engineering heritage: Northern England.* Thomas Telford, London, 1996, 265–266 (HEW 976).

2. LEAD P. *The Caldon Canal and tramroads.* Oakwood Press, Oxford, 1990.

14. Harecastle Tunnels

HEW 54
SJ 849 517 to SJ 837 541

The original Harecastle Tunnel on the Trent and Mersey Canal was engineered by one of the pioneer canal engineers, James Brindley, and its 1 mile 1120 yd length took nine years to construct between 1766 and 1775. It was the first tunnel in Great Britain to be constructed solely for transport purposes and even now ranks among the longest canal tunnels built. Fifteen working shafts were used in its construction, enabling work to be carried out at 32 different faces. Many problems were encountered during the driving of the tunnel, not the least of which was the ingress of large quantities of water. Eventually steam pumps were used to clear the workings.

This tunnel was only 8 ft 6 in. wide and did not have a towpath, so that boats had to be laboriously 'legged' through. The restricted width prevented boats from passing simultaneously in both directions, which caused a severe bottleneck on the canal as the traffic built up. Eventually a decision was made to build a new tunnel alongside the old one.

HEW 465
SJ 849 517 to SJ 837 541

Harecastle New Tunnel, 1 mile 1166 yd long, is slightly longer than its predecessor and lies to the east. Engineered by Thomas Telford with James Potter as Resident

South Portal, Harecastle New Tunnel

R. CRAGG

Engineer and Daniel Pritchard and William Hoof as contractors,[1] it was built in two years and opened in 1827, an indication of the progress made in tunnelling in the intervening 52 years.

The New Tunnel is wider and had a towpath. Both tunnels were used for a time, carrying traffic in opposite directions, but the Old Tunnel was closed in 1918 and is now derelict largely as a result of mining subsidence. The New Tunnel remains in use, although the size of boats which may pass through is limited by subsidence within the tunnel. The towpath is now impassable. A ventilation system has been installed at the south end of the tunnel.

1. LINDSAY J. *The Trent and Mersey Canal*. David and Charles, Newton Abbot, 1979, 105–110.

15. Hazlehurst (Denford) Aqueduct

HEW 554
SJ 954 536

Flyovers are now commonplace on motorways and have been in use on railways for many decades. They are also to be found on canals, as witness the Hazlehurst 'flyover' on the Leek branch of the Caldon Canal in Staffordshire. This aqueduct was necessary because of a realignment in

Hazlehurst Aqueduct

R. CRAGG

the local canal system on the opening of the Stoke to Leek railway in 1841.

The branch canal to Leek diverges slightly to the south and then for 660 yd runs parallel to the parent canal which, over this distance, has been lowered by about 26 ft by a flight of three locks. At this point the branch bends to the north-east and crosses over the main canal by means of an attractive aqueduct. The semicircular skew brick arch spans about 27 ft.

There is a second aqueduct, just to the north along the Leek branch canal, which carries the canal over the track of the railway.

Another attractive feature of this site is the cast-iron roving bridge[1] at SJ 948 537, which carries the towpath of the Leek branch over the head of Hollinshurst Top Lock. It has a semi-elliptical arch of 24 ft span.

1. Hazlehurst Iron Bridge (HEW 857) SJ 947 538.

16. Cheddleton Mills

HEW 1133
SJ 973 526

On the banks of the River Churnet at Cheddleton in Staffordshire is an interesting pair of water-wheel-driven flint grinding mills. Ground flint is used extensively in the nearby pottery industry.

When the mills were built is not known, but there are records of a corn mill on the site from as early as the thirteenth century. The older, the South Mill, was at some time converted from corn to flint grinding. The North Mill was built specifically for the flint trade and may have been the work of James Brindley, the pioneer canal engineer who was a millwright by trade. There is a mill by Brindley, dated 1752, at Leek.

The low breast wheels are 20 ft 5 in. (South Mill) and 22 ft (North Mill) in diameter. They are built from cast-iron segmental rims with oak spokes connecting the rims to cast-iron hubs, which are mounted on cast-iron shafts. The shaft of the North Mill is cruciform in section, 15¼ in. in size, whilst the South Mill has a hexagonal shaft 16 in. across the flats. The motion of each wheel is transmitted to the vertical driving shaft by two great bevelled gear wheels—one 10 ft in diameter and one 8 ft 6 in. diameter in the North Mill, and both 6 ft 6 in. diameter in the South

R. CRAGG

Cheddleton Mills, with the North Mill on the left

Mill. In addition to driving the grinding pan, the power of the wheel is also used to drive a hoist mechanism and pumps.

The mills have both been restored and are now maintained in working order by a Trust.

17. Froghall Basin

HEW 1590
SK 027 477

At Froghall, north of Cheadle in Staffordshire, is to be found a complex of canal wharves, lime kilns and associated tramways which were built by the Caldon Canal Company to facilitate the transport of limestone from the extensive quarries at Caldon Low to the north-east.

Construction of the Caldon Canal from its junction with the Grand Trunk (Trent and Mersey) Canal began in 1776 under the direction of Hugh Henshall, its Engineer. By about 1779 it had been completed to a terminus near Froghall Bridge, south of the present basin. From this point a tramway was built to the north-east, climbing some 700 ft in a length of about 2½ miles to the quarries. This tramway was not a great success and within a year was being reported as being 'very crooked, steep and uneven in its degree of declivity'.[1] To overcome these problems a section of the tramway in the vicinity of Shirley Hollow (SK 044 485) was rebuilt on a new alignment. At the same time, under an Act of 1783, the canal was extended a further 540 yd through a short tunnel to its present terminus. Even this improved alignment did not apparently prove to be satisfactory and by 1785 a new tramway was built, this time running to the south of the old route. In 1801 John Rennie advised the Company on the route of a third tramway and this was completed in 1804, running further to the south, just north of the village of Whiston. Traces of the route of this tramway and its inclined planes can still be seen.

Between 1807 and 1811 the canal was extended (as the Uttoxeter Canal) to the south-east and traces of the first lock on this canal can be seen to the west of the road through the site.

Although Rennie's tramway operated successfully for over 30 years, breakages of the rails and other problems led the company to ask its Engineer, James Trubshaw, to design a fourth, and last, route for the tramway which was opened in July 1847. Built to a gauge of 3 ft 6 in., the tramway included a 330 yd tunnel and was worked by cable and gravity. Many traces of this route can be seen,

Froghall Basin

R. CRAGG

both on the map and on the ground, cutting an almost straight alignment across the country.

Present-day Froghall Basin is a pleasant picnic site but old photographs and the traces of many buildings around the site reveal that it was once a centre of activity: there were extensive limekilns, many of which can still be seen today, and the basin was busy with the transshipment of limestone from the tramway wagons to the waiting canal boats. Surviving buildings on the site include a two-storey canal-side warehouse with a crane housing projecting over the canal, and a stable building.

The Caldon Canal is still open and used by pleasure craft although the Uttoxeter Canal was used for the route of a railway in the 1840s. The last tramway closed in 1920, the quarries having been reached by a standard gauge railway from the north-west in 1905.

1. LEAD P. *The Caldon Canal and tramroads*. Oakwood Press, Oxford, 1990, 69.

18. Macclesfield Canal

HEW 1136
SJ 836 546 to
SJ 961 884

The Macclesfield Canal was born too late, right at the end of the canal era. Its route was surveyed by Thomas Telford early in 1825, and the enabling Act received Royal Assent in April 1827. It was completed in 1831, at the beginning of the railway era, and like the Birmingham and Liverpool Junction Canal, was designed not to meander along the contours of the land but to take a straighter course on embankments. Nevertheless, it was never able to take full advantage of its 'modern' design, partially because of the jealousies of existing canal companies, particularly the Trent and Mersey. The Engineer, Thomas Brown, like Telford, appears to have played little part in its construction, the works being supervised by William Crosley with Daniel Pritchard and William Hoof as contractors for the length to Hall Green.

The canal provided a link from the Potteries in the south to Manchester in the north, joining the Trent and Mersey Canal at a junction at Hall Green near Kidsgrove, and the Peak Forest Canal above the top lock at Marple. Roughly following the 400 ft contour, the many embankments involved heavy earthworks. There is a substantial aqueduct at Bollington (SJ 934 779). The rise in the canal was concentrated in one flight of twelve locks at Bosley.[1] The stop lock at Hall Green is a relic of the jealousies between the various canal companies. It adjusts the level of the water by only 1 ft in order to ensure that no water passed from the summit level of the Trent and Mersey into the Macclesfield.

1. Bosley Locks (HEW 1314) SJ 906 663.

19. Jodrell Bank Radio Telescope

HEW 471
SJ 795 711

The University of Manchester's radio telescope at Jodrell Bank is a familiar feature of the Cheshire landscape. The need for a fully steerable instrument of such a size developed as a result of the limitations of the 218 ft transit telescope, built in 1947, which was its predecessor. The telescope was designed by Husband and Company of Sheffield to meet the exacting requirements of the astronomer Professor (later Sir) Bernard Lovell.[1]

Thirty different firms were engaged on the construction, which was completed in 1957.[2] A paraboloid bowl 250 ft in diameter, formed of welded steel sheets, is carried on a space frame of structural steel. Radio signals from space are reflected to the focus of the bowl, where aerials are mounted on a steel mast 62 ft 6 in. high. The bowl and its supporting structure are carried by trunnion bearings, which were formerly part of the 15 in. gun turrets of HMS *Revenge* and HMS *Royal Sovereign*.

The two supporting towers, which carry the bowl assembly at an elevation of 165 ft, are braced to form a yoke under the bowl and are mounted on bogies on a track of two concentric rails with an overall diameter of 353 ft. The telescope can thus be rotated to point at any portion of the sky.

The moving parts weigh approximately 2000 tons and the whole structure nearly 3200 tons, following modifications made in 1970. The loads on the trunnion bearings and the rail tracks were then reduced by the incorporation of two semicircular wheel girders under the bowl, which transferred one-third of the load to a new central rail track.

There is a museum and a planetarium on the site, but

HUSBAND & CO.

Jodrell Bank Radio Telescope

the telescopes and their control rooms are not open to the general public.

1. HUSBAND H. C. The Jodrell Bank radio telescope. *Proc. Instn Civ. Engrs*, 1958, **9**, 65–86.

2. LOVELL Sir B. *The Story of Jodrell Bank*. Oxford University Press, London, 1968.

20. Weaver Navigation

HEW 1191
SJ 655 664 to SJ 493 818

The River Weaver flows northwards from central Cheshire through Nantwich, Winsford and Northwich to join the Manchester Ship Canal at Weston Point, near Runcorn.

The last 7 miles of the river were always navigable and substantial improvements in the 1730s enabled 40 ton craft to reach Winsford, 20 miles upstream, the Navigation passing through eleven locks *en route*. The main traffic was coal and salt. Problems with the tidal entrance to the Mersey were overcome by Thomas Telford in 1810, when the 4 mile Weston Canal and a new basin and river lock were opened at Weston Point. Improvements recommended by William Cubitt were made in the 1840s to enable the Navigation to be used by 100 ton flats. Enlarged locks designed by E. L. Williams Junior enabled coastal traffic to use the Navigation in 1870.

At the end of the nineteenth century a new dock was built at Weston Point and a number of the smaller locks were replaced by fewer and larger locks, 229 ft by 42 ft 6 in., with gates operated by Pelton wheels. The construction in 1875 of a boat lift at Anderton enabled interchange of traffic with the Trent and Mersey Canal and in 1894 the opening of the Manchester Ship Canal[1] gave a new outlet to deeper water at Eastham.

Other interesting structures associated with the Weaver Navigation are Dutton Horse Bridge,[2] the railway viaducts at Frodsham,[3] and several swing bridges including Acton Bridge,[4] Sutton Weaver Bridge and Town and Hayhurst (Navigation)[5] at Northwich. These swing bridges, designed by J. A. Saner, have the unusual feature of being supported partly on piles and partly on floating caissons, to counteract subsidence from salt mine workings.[6] The locks at Sutton (SJ 586 769), Hunts (SJ 656 729) and Vale Royal (SJ 640 704), constructed in 1874–87, were operated by turbines, designed by J. W. Sandeman and F. G. M. Stoney on Schiele's principle using 18 in. diameter wheels with 20 double buckets 2 in. deep. Most of the large locks on the Weaver are now operated electro-hydraulically.

Commercial traffic to Winsford ceased in the 1950s but coasters up to 800 tons still navigate as far as Northwich.

1. RENNISON R. W. *Civil engineering heritage: Northern England.* Thomas Telford, London, 1996, 263–265 (HEW 88).

2. Dutton Horse Bridge (HEW 1316) SJ 584 767.

3. Frodsham Viaduct (HEW 1168) SJ 529 786 and SJ 534 790.

4. Acton Bridge (HEW 1281) SJ 601 761.

5. Town and Hayhurst Bridges (HEW 1282) SJ 656 736.

6. SANER J. A. Swing bridges over the River Weaver at Northwich. *Min. Proc. Instn Civ. Engrs*, 1899–1900, **140**, 72–108.

21. Anderton Boat Lift

HEW 286
SJ 647 752

This lift, which is unique in Britain, transfers canal boats between the River Weaver Navigation and the Trent and Mersey Canal through a 50 ft 4 in. difference in water level.[1,2] At the top level the lift is linked to the canal by means of a 162 ft 6 in. long aqueduct which has twin channels.

The lift was originally operated by water pressure; the two caissons were supported on 3 ft diameter rams working in the cylinders of the hydraulic presses, which were linked through valves. The caissons are of wrought iron, being 75 ft long and 15 ft 6 in. wide, with gates at each end. By 1904 the lift, which was opened in 1875, needed overhauling and the main rams required renewing. It was therefore converted to a counterbalanced structure driven by electric motors, with the caissons suspended from wire ropes carrying cast-iron weights, some 252 tons to each caisson.[3] It has been disused since 1982.

The lift, which is constructed of wrought and cast iron, was designed by Sir Edward Leader Williams, Edwin Clark and J. W. Sanderson and was the forerunner of several similar lifts constructed in Europe. It is now a scheduled ancient monument and there are currently plans to restore it to working order.[4]

1. DUER S. Hydraulic canal lift at Anderton, on the River Weaver. *Min. Proc. Instn Civ. Engrs*, 1875–76, **45**, 107–129.

2. WELLS L. B. Correspondence: Some canal, river and other works in France, Belgium and Germany, by L. F. Vernon Harcourt. *Min. Proc. Instn Civ. Engrs*, 1888–89, **96**, 223–226.

3. Reconstruction of the Anderton boat lift. *Engineer*, 1908, **106**, 24 July, 82–84, 92.

4. HAYWARD D. Raising expectations. *New Civil Engineer*, 1996, 29 Feb., 16–18.

BRITISH WATERWAYS BOARD ARCHIVE

Anderton Lift

22. Grand Junction Railway

HEW 1129
SJ 800 402 to
SJ 578 951

The Grand Junction Railway, introduced in Chapter 6, was the fruit of the experience of George Stephenson and the young men about him in perfecting their craft of railway construction, learnt in the building of the Liverpool and Manchester Railway[1] and in maintaining and improving that pioneering work.

Across the area covered by this chapter, the general alignment of the railway was north-west, passing through Crewe and entering the valley of the River Weaver at Northwich. From Northwich the line followed the Weaver before turning north, and finally north-east, to its junction with the Liverpool and Manchester Railway at Newton-le-Willows.

In common with most contemporary railways, the Grand Junction selected a location for its main locomotive depot about halfway along its length, other examples being the selection of Wolverton by the London and Birmingham Railway and Swindon by the Great Western. The site chosen by the Grand Junction was near the small village of Monks Coppenhall but the name adopted was Crewe, after the nearby Crewe Hall. Begun in 1842, Crewe Locomotive Works[2,3] expanded rapidly and in its heyday in the 1880s extended for about 2 miles. Today, little remains on site to remind us that the works not only manufactured and maintained locomotives, carriages and wagons but also bridges, rails, points and crossings and all the hardware associated with the operation of a large railway.

Crewe station became a major interchange point with lines eventually opening to Shrewsbury, Chester, Manchester and Stoke-on-Trent. The station was extensively remodelled in 1985.

North of Crewe, the Grand Junction Railway Company built some notable viaducts such as those at Vale Royal (SJ 643 706), with five arches of 62 ft span, Dutton (see below) and the twelve-arch structure over the Mersey south of Warrington at SJ 600 866. This last bridge ceased to serve the main line when the Manchester Ship Canal was built and the railway was diverted to a new high-level bridge at Acton Grange.

1. RENNISON R. W. *Civil engineering heritage: Northern England*. Thomas Telford, London, 1996, 249–255 (HEW 223).

2. Crewe Locomotive Works (HEW 276).

3. REED B. *Crewe locomotive works and its men*. David and Charles, Newton Abbot, 1982.

23. Dutton Viaduct

HEW 154
SJ 582 764

Dutton Viaduct was designed by Joseph Locke to carry the Grand Junction Railway across the valley of the River Weaver. It was built between 1834 and 1836 at a cost of £54 440 and opened in 1837. The contractor was David McIntosh with Edward Betts as Agent. Of masonry construction, the viaduct consists of 20 segmental arches of 60 ft span with a rise of 17 ft 6 in. Rail level is at a maximum height of 65 ft.

Nearby is to be found Dutton Horse Bridge[1,2] which carries the towpath of the River Weaver across a sluice channel. Built in 1919 to the design of John Saner, Engineer to the River Weaver Navigation Trust, the bridge has two 96 ft spans each with a pair of laminated timber arches.

1. Dutton Horse Bridge (HEW 1316) SJ 584 767.

2. Winney M. Stop the rot. *New Civil Engineer*. 1994, 26 May, 16–17.

24. Acton Grange Viaduct

HEW 1229
SJ 590 857

The Manchester Ship Canal[1] has many bridges across it, from the big railway viaduct at Runcorn[2] in the west to an insignificant small bridge at the Manchester end. Several are swing bridges and many of the bridges have the original (1893) superstructures.

Apart from the Barton Aqueduct,[3] perhaps the most notable is the high-level fixed span carrying the West Coast Main Line to the north. An opening span would not have been suitable, so the line of the then London and North Western Railway (L&NWR), originally the Grand Junction Railway, was diverted to the west over a considerable distance to enable it to climb at an acceptable gradient to clear the Ship Canal above the mast height of ships. The bridge was designed by William Bean Farr, then head of the bridge office of the L&NWR.

A branch line to Chester diverges at this point and this also had to be carried on the new viaduct, which is on a very considerable skew; the skew span, between masonry abutments, is 263 ft to provide a square span of only 120 ft.

The type of structure chosen was a form of lattice

Acton Grange Viaduct in 1904

SCIENCE MUSEUM PICTURE LIBRARY

girder known as a Whipple-Murphy truss, very popular for a period around 1890–1910. There are five main girders.

1. RENNISON R. W. *Civil engineering heritage: Northern England.* Thomas Telford, London, 1996, 263–265 (HEW 88).

2. Ibid. 261–262 (HEW 196).

3. Ibid. 266–267 (HEW 28).

25. Warrington Bridge

HEW 254
SJ 608 879

This bridge, which carries the A49 road, was built in 1915 and is the sixth bridge to occupy the same site at this important crossing of the River Mersey, which is still tidal at this point. A bridge was recorded here as early as 1305.[1]

The bridge has a clear span of 134 ft and is 80 ft wide between the parapets. Eight reinforced concrete parabolic arch ribs carry the deck, with a shallow rise to span ratio of 1 to 10, dictated by the level of the road. The lateral thrusts of 350 tons at the abutments are taken by reinforced concrete counterforts at each rib, bearing on inclined reinforced concrete piles which are driven into the hard clay on the line of the resultant thrust.

Temporary concrete hinges were used to accommodate 3 in. of settlement that took place during the construction period. This was perhaps the first bridge to be built in Britain with reinforced concrete hinges, which were located at the crowns of the arch ribs and reinforced by shear bars and interlocking spirals of reinforcement. The thrust on these hinges was about 240 tons.

The deck loading is transferred to the ribs by means of short columns; the bridge was subjected to careful monitored test loading using a combination of tramcars and steam rollers before it was opened. At high water the springings of the arch ribs are under water, which gives some idea of the physical constraints of the site which faced the designers, Webster and Fitzsimons, who acted in conjunction with Considère Constructions Ltd. In addition, so as not to interfere with road traffic, the bridge had to be built in two halves. It was constructed by Alfred Thorne and Sons.

1. SCOTT W. L. *Reinforced concrete bridges*. Crosby Lockwood, London, 1931, 231–232.

26. Runcorn–Widnes Road Bridge

HEW 1063
SJ 510 835

The main span of 1082 ft of this elegant modern bridge, which crosses the River Mersey with a clearance of 75 ft, is longer than that of any other steel arch in Britain. It is twice the span of the Tyne Bridge at Newcastle,[1] which was the longest at the time of its opening in 1928.

Mott, Hay and Anderson designed the bridge to replace the former transporter bridge, built in 1905, which was the longest (1000 ft span) and the lowest (75 ft) of the five transporter bridges built in this country.[2] A two-pinned bowstring arch design was adopted after wind tunnel tests indicated that the presence of the adjacent railway bridge could, under certain wind conditions, produce severe oscillations if a conventional suspension bridge design were to be used.

Leonard Fairclough Ltd and Dorman Long Ltd began construction in April 1956 and the bridge was opened in July 1961.

The bridge has side spans of 250 ft, one crossing the Manchester Ship Canal with a clearance of 80 ft. The road was widened between 1975 and 1977 to four lanes, with a single footway cantilevered off the side of the road deck.

CHESHIRE COUNTY COUNCIL

Runcorn–Widnes Road Bridge, with the Railway Bridge beyond

1. RENNISON R. W. *Civil engineering heritage: Northern England*. Thomas Telford, London, 1996, 46–47 (HEW 91).

2. ANDERSON J. K. Runcorn–Widnes Bridge. *Proc. Instn Civ. Engrs*, 1964, **29**, 535–570.

27. Runcorn Railway Bridge

HEW 196
SJ 509 835

This important structure was built in 1868 to carry the London and North Western Railway across the River Mersey. Hitherto the route from London to Liverpool was via Warrington, Earlestown and the Liverpool and Manchester Railway.

The river is crossed by three main spans of 305 ft formed by wrought iron double-web lattice girders supported on stone piers. The overhead lattice bracing to the top flange is masked by stone portals which, perhaps unfortunately, are surmounted by castellated turrets in the Victorian Gothic revival style reaching to 50 ft above rail level. The bridge carries the railway 75 ft above the river to allow sufficient headroom for ships passing beneath. In order to achieve such a height in the approach to the bridge, the railway is carried on 59 brick arches on its rise to the Widnes end of the bridge.

The Engineer was William Baker, the contractor Thomas Brassey and the ironwork was supplied by J. Cochrane and Sons.

28. Manchester Airport

HEW 1963
SJ 819 851

Located 8 miles south of the city centre and owned by the local authorities, Ringway Airport developed from 1935 onwards with a remarkable record of almost continual growth. The first terminal, control tower, hangar and the general layout of the airfield came into being in 1938.[1] The main runway was extended progressively to its present length of 10 000 ft, which required the diversion of the River Bollin,[2] and there are plans to construct a second, parallel, main runway. A new terminal was built in 1962 and a second terminal was constructed in 1993, when a new rail link and station were also provided.[3,4,5]

Manchester Airport was designed in phases by successive City Engineers for Manchester and is now the second busiest airport in the country after London Heathrow.[6]

1. HAYES J. and BAMBER J. P. The Manchester Airport. *Proc. Instn Civ. Engrs*, 1962, **23**, 213–224.

2. NAYLOR A. E. Improvement of the runway at Manchester Airport. *Proc. Instn Civ. Engrs*, 1984, **76**, 221–236.

3. CONGDON L. and TAYLOR A. Expansion of Manchester Airport. *Proc. Instn Civ. Engrs, Transport*, 1992, **95**, May, 79–86.

4. TAYLOR A. and DUFFY P. S. Manchester Airport terminal 2: project planning and implementation. *Proc. Instn Civ. Engrs, Civ. Engng*, 1993, **97**, May, 56.

5. EDGE D. J. and HEATHER M. J. Manchester Airport terminal 2: terminal infrastructure. *Proc. Instn Civ. Engrs, Transport*, 1994, **105**, Feb., 21.

6. RUSSELL L. Expanding on all fronts. *New Civil Engineer*, 1996, 25 July, xii–xiii.

Additional Sites

Numbers in bold type indicate Historical Engineering Works (HEW) site numbers.

1. North Wales

Aberffraw Bridge, **163**, SH 355 689

Bangor Pier, **427**, SH 584 732

Corwen Bridge, **1646**, SJ 067 432

Dinorwig Tramroad and Padarn Railway, **1286**, SH 526 678 to SH 59 60

Holyhead Road–Glyn Diffws Pass, **1505**, SH 995 443 to SJ 991 444

Llangollen Chain Bridge, **727**, SJ 199 432

Maentwrog Dam, **399**, SH 673 377

Pont Cilan, River Dee, **1644**, SJ 021 375

Pont Dyfrydwy, River Dee, **1645**, SJ 052 412

Rhos on Sea Offshore Breakwater, **1431**, SH 845 807

2. Mid-Wales

Berriew Aqueduct, **1465**, SJ 189 006

Brithdir Aqueduct, **522**, SJ 198 022

Buttington Bridge, **852**, SJ 246 089

Claerwen Dam, **186**, SN 870 636

Vyrnwy Canal Aqueduct, **1112**, SJ 254 196

Welshpool Canal Aqueduct, **1836**, SJ 227 074

3. South Wales

Bridgend Old Town Bridge, **1031**, SS 904 798

Bryn Tramroad Bridge, near Port Talbot, **1045**, SS 787 923

Cenarth Bridge, **166**, SN 269 415

Leckwith Ancient Bridge, **161**, ST 851 915

Llwchwr Road Viaduct, **1231**, SS 562 981

Maesteg Bridge, **525**, SS 851 915

Merthyr (Ynysgau) Bridge, **805**, SO 047 062

Pont Rhyd y Fen Bridge, **1004**, SS 796 942

Pont Spwdwr Road Bridge, **1218**, SN 434 059

Resolven Aqueduct, **2082**, SN 830 027

4. The Bristol Area, Bath and North Wiltshire

Almondsbury Motorway Interchange, **49**, ST 617 837

Ashton Swing Bridge, Bristol, **1535**, ST 568 721

Blagdon Pumping Station, **1538**, ST 503 600

Bradford on Avon Rail Tunnel, **382**, ST 825 606

Bradford on Avon Town Bridge, **1514**, ST 826 609

Bradford on Avon Wharf, Kennet and Avon Canal, **383**, ST 825 601

Brislington Road Bridge, Bristol, **1143**, ST 617 727

Bristol Docks Hydraulic System, **1536**, ST 572 723

Cheltenham Road Railway Bridge, Bristol, **99**, ST 589 746

Cleveland Bridge, Bath, **650**, ST 753 657

Limpley Stoke Viaduct, **377**, ST 781 620

Mauds Heath Causeway, near Chippenham, **680**, ST 972 737 to ST 919 739

North Parade Bridge, Bath, **339**, ST 754 647

Prince's Street Swing Bridge, Bristol, **1565**, ST 586 723

Priston Mill, **1209**, ST 695 615

Severn Cable Tunnel, **1144**, ST 555 902

Victoria Pumping Station, Bristol, **1537**, ST 577 737

5. Gloucestershire, Hereford and Worcester

Berkeley Nuclear Power Station, **1181**, ST 660 995

Bibury Bridge, **1140**, SP 115 068

Bringewood Forge Bridge, **1278**, SO 454 750

Bullo Pill, **1332**, SO 690 099

Danzey Green Post Mill (at Avoncroft Museum of Buildings), **864**, SO 454 750

Donnington Brewery, near Stow-on-the-Wold, **1271**, SP 174 272

Droitwich Canal, **183**, SO 842 599 to SO 904 635 to SO 422 629

Evesham Bridge, **1700**, SP 034 431

Foregate Street Bridge, Worcester, **1912**, SO 849 552

Gloucester Lock Bridge, **1525**, SO 827 184

Hewlett's Reservoirs, Cheltenham, **1308**, SO 97 22

Leintwardine Bridge, **1906**, SO 404 738

Masonry Bridge, Cinderford, **1311**, SO 642 125

Monkhide Bridge, near Hereford, **1874**, SO 611 440

New Mills Viaduct, Ledbury, **1987**, SO 702 388

Pont ar Ynys Bridge, **1979**, SO 327 287

Rainbow Hill Bridge, Worcester, **1911**, SO 854 555

Redbrook Incline, Monmouth Tramroad, **627**, SO 537 102

River Teme Railway Bridge, near Ludlow, **2053**, SO 517 717

Stourport Bridge, **1051**, SO 808 711

Tenbury Bridge, **1907**, SO 595 686

Tintern Tramway Bridge, **930**, SO 530 003

Wergins Bridge, nr Hereford, **1960**, SO 529 447

Witcombe Reservoir, near Gloucester, **1309**, SO 900 155

6. West Midlands and South Staffordshire

Anker Viaduct, Tamworth, **886**, SK 213 036

Birchills Canal Aqueduct, **830**, SK 009 004

Brownhills Canal Aqueduct, **283**, SK 053 064

Farmers Bridge Locks, Birmingham, **1910**, SP 065 873

Holliday Street Aqueduct, Birmingham, **1651**, SP 064 865

Lee Bridge, Birmingham, **1704**, SP 047 877

Rotton Park Reservoir, **1584**, SP 043 867

Spon Lane Locks, Birmingham, **1964**, SO 999 899

Taylor's Aqueduct, Birmingham, **1909**, SP 018 949

Trent Aqueduct, Rugeley, **1719**, SK 039 195

Winson Green Bridge, Birmingham, **1705**, SP 043 878

7. Shropshire

Atcham Old Bridge, near Shrewsbury, **732**, SJ 541 093

Boreton Cast Iron Bridge, Cound Brook, **987**, SJ 517 068

Borle Brook Footbridge, **977**, SO 753 817

Bridgnorth Station, **640**, SO 715 926

Greyfriars Footbridge, Shrewsbury, **1104**, SJ 495 121

Hadley Park Lock, Shrewsbury Canal,Telford, **938**, SJ 672 132

Ludford Bridge, Ludlow, **2052**, SO 512 742

Norbury Junction Canal Bridge, **462**, SJ 793 228

Old Forge Bridge, Hampton Loade, **971**, SO 747 863

Porthill Bridge, Shrewsbury, **1102**, SJ 485 126

Puleston Bridge, nr Newport, **1959**, SJ 734 219

8. Cheshire and North Staffordshire

Bollinhirst Dam, **1151**, SJ 973 836

Bottoms Dam, **1153**, SJ 945 716

Congleton Viaduct, **1336**, SJ 877 627

Crewe Station, **1497**, SJ 711 548

Dane Aqueduct, Macclesfield Canal, **1343**, SJ 906 652

Dock Road Edwardian Pumping Station, Northwich, **1939**, SJ 658 733

Horse Coppice Dam, **1152**, SJ 968 837

Lamberts Lane Bridge, Macclesfield Canal, **2040**, SJ 864 620

Moving of the Old Academy Building, Warrington, **1085**, SJ 607 879

North Rode Viaduct, **519**, SJ 896 657

Reinforced Concrete Arch Bridges, Runcorn Area, **259**, SJ 533 797 and SJ 554 797

Queensferry Bridge, **1317**, SJ 322 687

Rudyard Dam, near Leek, **1472**, SJ 951 583

Stoke-on-Trent Station Roof, **764**, SJ 879 456

Stone Railway Station, **1880**, SJ 897 345

Weaver Navigation Water Powered Machinery, **1339**, SJ 586 769, SJ 656 729 and SJ 640 704

General Bibliography

Abbott W. *The turnpike road system in England and Wales 1663–1840*. Cambridge University Press, 1972.

Adamson S. H. *Seaside piers*. Batsford, London, 1977.

Balkwill R. and Marshall J. *The Guinness book of railway facts and feats*. Guinness Publishing, London, 1993.

Barrie D. S. M. *A regional history of the railways of Great Britain, Volume 12: South Wales*. David and Charles, Newton Abbot, 1980.

Baughan P. E. *A regional history of the railways of Great Britain, Volume 11: North and Mid-Wales*. David and Charles, Newton Abbot, 1980.

Baughan P. E. *The Chester and Holyhead Railway*. David and Charles, Newton Abbot, 1972.

Beaver P. *A history of lighthouses*. Peter Davies, London, 1971.

Beaver P. *A history of tunnels*. Peter Davies, London, 1972.

Beckett D. *Brunel's Britain*. David and Charles, Newton Abbot, 1980.

Beckett D. *Stephenson's Britain*. David and Charles, Newton Abbot, 1984.

Berridge P. S. A. *The girder bridge*. Maxwell, London, 1969.

Biddle G. *Victorian stations*. David and Charles, Newton Abbot, 1973.

Biddle G. *The railway surveyors*. Ian Allan, London, 1990.

Binnie G. M. *Early dam builders in Britain*. Thomas Telford, London, 1987.

Binnie G. M. *Early Victorian water engineers*. Thomas Telford, London, 1981.

Blackwall A. *Historic bridges of Shropshire*. Shropshire Libraries, Shrewsbury, 1985.

Blower A. *British railway tunnels*. Ian Allan, London, 1964.

Boyd J. I. C. *Narrow gauge railways in mid-Wales (1850–1970)* Oakwood Press, Oxford, 1986.

Boyd J. I. C. *The Festiniog Railway*. Odhams Press, London, 1962.

British Bridges. Public Works, Roads and Transport Congress, London, 1933.

Broadbridge S. R. *The Birmingham Canal Navigations, Volume 1: 1768–1846*. David and Charles, Newton Abbot, 1974.

Brown R. A. *et al. The history of the King's works, Volume 1: The Middle Ages*. HMSO, London, 1963.

Buchanan R. A. *The engineers*. Jessica Kingsley, London, 1989.

Burton A. *The canal builders*. Eyre Methuen, London, 1972.

Burton A. *The railway builders*. John Murray, London, 1992.

Carter E. *A historical geography of the railways of the British Isles*. Cassell, London, 1959.

Cartwright R. and Russell R. T. *The Welshpool and Llanfair Light Railway*. David and Charles, Newton Abbot, 1989.

Chrimes M. M. *Civil engineering, 1839–1889: a photographic history*. Alan Sutton, Gloucester, 1991.

Christiansen R. *A regional history of the railways of Great Britain, Volume 7: The West Midlands*. David and Charles, Newton Abbot, 1973.

Christiansen R. *A regional history of the railways of Great Britain, Volume 13: Thames and Severn*. David and Charles, Newton Abbot, 1981.

Corlett E. G. B. *The iron ship*. Moonraker Press, Bradford-on-Avon, 1975.

Cossons N. and Trinder B. *The Iron Bridge*. Ironbridge Gorge Museum Trust and Moonraker Press, Bradford-on-Avon, 1979.

Cunliffe B. *Roman Bath discovered*. Routledge and Kegan Paul, London, 1984.

Fowler C. E. *The ideals of engineering architecture*. Spon, London, 1929.

Gorvett D. *Bridges over the River Wye*. C. J. and A. Bryant, Whitney, 1984.

Hadfield C. *The canals of South Wales and the Border*. David and Charles, Newton Abbot, 1967.

Hadfield C. *The canals of the West Midlands* David and Charles, Newton Abbot, 1985.

Hadfield C. and Skempton A. W. *William Jessop, engineer*. David and Charles, Newton Abbot, 1979.

Handford M. *The Stroudwater Canal*. Alan Sutton, Gloucester, 1979.

Helps Sir A. *Life and labours of Mr Brassey*. Evelyn Adams and Mackay, London, 1969 (first published 1872).

Heyman J. *The masonry arch*. Ellis Horwood, Chichester, 1982.

Hopkins H. J. *A span of bridges*. David and Charles, Newton Abbot, 1970.

Household H. *The Thames and Severn Canal*. Alan Sutton, Gloucester, 1983.

Hughes S. *The archaeology of the Montgomeryshire Canal*. Royal Commission on Ancient and Historical Monuments in Wales, Aberystwyth, 1988.

Jervoise E. *The ancient bridges of Wales and Western England*. Architectural Press, London, 1936; republished by E.P. Publishing, East Ardsley, Wakefield, 1976.

Labrum E. A. *Civil engineering heritage: Eastern and Central England*. Thomas Telford, London, 1994.

Lead P. *Agents of revolution: John and Thomas Gilbert—entrepreneurs*. University of Keele, Keele, 1989.

Lead P. *The Caldon Canal and Tramroads*. Oakwood Press, Oxford, 1990.

Lindsay J. *The Trent and Mersey Canal*. David and Charles, Newton Abbot, 1979.

Long P. J. and Awdry Revd W. V. *The Birmingham and Gloucester Railway*. Alan Sutton, Gloucester, 1987.

Lovell Sir B. *The story of Jodrell Bank*. Oxford University Press, London, 1968.

Macdermot E. T. (revised Clinker C. R.) *History of the Great Western Railway, Volume 1: 1833–1863; Volume 2: 1863–1921*. Ian Allan, London, 1982.

Margary J. D. *Roman roads in Britain*. John Baker, London, 1973.

Marshall J. *A biographical dictionary of railway engineers*. David and Charles, Newton Abbot, 1978.

Morriss R. K. *Canals of Shropshire*. Shropshire Books, Shrewsbury, 1991.

Otter R. A. *Civil engineering heritage: Southern England*. Thomas Telford, London, 1994.

Penfold A. (ed.) *Thomas Telford: engineer*. Thomas Telford, London, 1980.

Phillips G. *Thames crossings, tunnels and ferries*. David and Charles, Newton Abbot, 1981.

Pugsley Sir A. (ed.) *The works of Isambard Kingdom Brunel*. Institution of Civil Engineers and University of Bristol, 1976.

Reader W. J. *Macadam: The McAdam family and the turnpike roads 1798–1861*. Heinemann, London, 1980.

Reed B. *Crewe locomotive works and its men*. David and Charles, Newton Abbot, 1982.

Rennison R. W. *Civil engineering heritage: Northern England*. Thomas Telford, London, 1996.

Rolt L. T. C. *George and Robert Stephenson*. Penguin Books, Harmondsworth, 1978.

Rolt L. T. C. *Isambard Kingdom Brunel*. Penguin Books, Harmondsworth, 1989.

Rolt L. T. C. *Railway adventure*. David and Charles, Newton Abbot, 1970.

Rolt L. T. C. *Thomas Telford*. Penguin Books, Harmondsworth, 1979.

Roscoe T. *The book of the Grand Junction Railway*. London, 1839.

Ruddock E. C. *Arch bridges and their builders, 1735–1835*. Cambridge University Press, 1979.

Russell R. *Lost canals of England and Wales*. David and Charles, Newton Abbot, 1971.

Skempton A. W. *British civil engineering 1640–1840: A bibliography of contemporary printed reports, plans and books*. Mansell Publishing, London, 1987.

Smiles S. *The lives of the engineers*. Murray, London, 1862 (and subsequent editions).

Steel W. L. *The history of the London and North Western Railway*. Railway and Travel Monthly, London, 1914.

Tann J. *Gloucestershire woollen mills*. David and Charles, Newton Abbot, 1967.

Telford T. (ed. Rickman J.) *Life of Thomas Telford, with a folio atlas of copper plates*. London, 1838.

Trinder B. *The making of the industrial landscape*. Dent, London, 1982.

Whishaw F. *The railways of Great Britain and Ireland*. Weale, London, 1842; reprinted by David and Charles, Newton Abbot, 1969.

Yeomans D. T. *The trussed roof: its history and development*. Scolar Press, Aldershot, 1992.

Name Index

* indicates a Past President of the Institution of Civil Engineers

Engineers

Abt, Dr Roman (1850–1933), 31
Alexander, Daniel Asher (1768–1846), 6
Arnodin, Ferdinand Joseph (1845–1924), 108
Bage, Charles (1752–1822), 244
*Baker, Sir Benjamin (1840–1907), 11
Baker, William (1819–78), 244, 287
*Barlow, William Henry (1812–1902), 120
*Barry, Sir John Wolfe (1836–1918), 103
*Bateman, John Frederick La Trobe (1810–89), 80, 267
Berridge, Percy Stuart Attwood (1901–80), 93
*Bidder, George Parker (1806–78), 211, 212
Binnie & Partners, 30, 66
Binnie, Deacon and Gourlay, 78
Birch, Eugenius (1818–84), 118
Brereton, Robert Pearson (circa 1815–94), 70, 136, 169
Brindley, James (1716–72), 185, 189, 195, 196, 215, 216, 217, 218, 219, 271, 273
Brown, Thomas (fl.1790–1832), 276
Brunel, Isambard Kingdom (1806–59), 24, 70, 81, 86, 113, 119, 122, 123, 125, 130, 132, 134, 135, 150, 169, 182
Brunel, Sir Marc Isambard (1769–1849), 136
*Brunlees, Sir James William (1816–92), 26
Capper, C.H., 180
Cartwright, Thomas (d. 1810), 185
Christiani & Nielsen A/S, 15
Clark, Edwin (1814–94), 280
Clowes, Josiah (1735–94), 167, 168, 192, 252, 253
Coignet, Edmund (1865–1915), 124
Collin A.J. (b. 1862), 52
Considère Constructions, 285
Conybeare, Henry (d. 1884), 90
Crosley, William (fl.1817–31), 185, 262, 276
*Cubitt, Sir William (1785–1861), 207, 279
Dadford, Thomas (Junior) (d. 1806), 76, 107
Darby, Abraham I (1619–1717), 227, 238
Darby, Abraham III (1750–89), 238, 239
Deacon, George Frederick (1843–1909), 79
Dredge, James (1794–1865), 65, 142, 148
Edwards, William (1719–89), 67, 87
Fairbairn, Sir William (1789–1874), 23
Farr, William Bean (1856–1922), 283
Fowler, Sir Henry (1870–1938), 183
*Fowler, Sir John (1817–98), 232, 240
*Fox, Sir Douglas (1840–1921), 31
Fox, Sir Francis (1818–1914), 18, 31
Freeman Fox & Partners, 34, 115
George, Watkin, 91
Gooch, Sir Daniel (1816–59), 130, 138
Gooch, Thomas (1803–83), 192, 211, 212
Grover, John William (1836–92), 117
Hartley, Jesse (1780–1860), 266
*Hawkshaw, Sir John (1811–91), 8, 120, 137
*Hawksley, Thomas (1807–93), 42
Haynes, Robert H. (1862–1910), 108
*Hayter, Harrison (1825–98), 8
Hennebique, François (1842–1921), 88
Henshall, Hugh (1734–1816), 219, 274
Hodgkinson, Prof. Eaton (1789–1861), 23
Howard, Thomas (circa 1820–95), 122, 123
Husband & Co., 24, 277
Jessop, Josias (1781–1826), 61, 251
Jessop, William (1745–1814), 48, 104, 121, 258

Engineers (continued)
Kennard, Thomas William (1825–93), 106
Kennedy & Donkin, 34
Kinderley, Nathaniel (1673–1742), 268
Lee, Hedworth (d. 1873), 16, 32
*Locke, Joseph (1805–60), 190, 209, 283
Mansergh, James (1834–1905), 63, 229
Marks, G. Croydon (later Baron Marks of Woolwich) (1858–1938), 235
Maynall & Littlewood, 27
McAdam, John L. (1756–1836), 136, 140, 214, 257
McConnell, James (1815–83), 183
McKerrow, Alexander (circa 1835–1915), 26
Moorsom, Capt. William Scarth (1804–63), 182
Mott, Hay and Anderson, 115, 285
Mouchel, Louis Gustave (1852–1908), 89, 187, 245
Mylne, Robert (1733–1811), 164
Overton, George (fl.1790–1822), 83
Penson, Thomas (circa 1790–1859), 60
Piercy, Benjamin (1827–88), 59
Piercy, Robert (1825–94), 59
Potter, James (1801–57), 271
Potter, Joseph (circa 1756–1842), 223, 224
Prestressed Concrete Company, 245
Provis, William Alexander (1792–1870), 11
Ramsbottom, John (1814–97), 17
Rastrick, John Urpeth (1780–1856), 111, 209
*Rendel, James Meadows (1799–1856), 8
Rennie, George (1791–1866), 266
Rennie, John (1761–1821), 7, 104, 146, 149, 152, 222, 275
Reynolds, William (1758–1803), 91, 237
Robertson, Henry (1816–88), 51, 63
Robinson, John (b. 1832), 97
Ross, Alexander Mackenzie (1805–62), 16
Sanderson, John Watt (1842–1927), 280
Saner, John Arthur (1864–1952), 279, 283
Sheasby, Thomas (fl.1780–99), 80
Shelford, Sir William (circa 1832–1905), 97
*Simpson, James (1799–1869), 108, 128
Skempton, Prof. Alec Westley (1914–), 79
Smith, William (d. 1911), 32
Spooner, Charles Easton (1818–89), 33
Spooner, James (1789–1856), 33
Spooner, James Swinton (b. 1816), 38
Stephenson, George (1781–1848), 194, 209, 281
*Stephenson, Robert (1803–59), 5, 14, 16, 21, 22, 23, 189, 192, 211, 212
Stuttle, William (fl.1810–24), 12, 262
Sutherland, Alexander, 90
Szlumper, Sir James (1834–1926), 57
Szlumper, William Weeks (b. 1868), 57
*Telford, Thomas (1757–1834), 5, 7, 9, 11, 12, 13, 14, 15, 22, 45, 47, 49, 91, 140, 146, 161, 164, 177, 181, 189, 195, 198, 200, 201, 206, 207, 214, 227, 228, 231, 233, 239, 242, 247, 249, 252, 253, 254, 257, 258, 260, 271, 276, 279
Travers Morgan & Partners, 15, 51
Trevithick, Richard (1771–1833), 83, 100
Upton, John (fl. circa 1815), 74
Ward, Arthur.Walburgh (1877–circa 1962), 248
Ward, Richard James (1817–81), 117
Warren, James (circa 1803–70), 106
Webster and Fitzsimons, 285
Webster, John James, 27
Weston, Samuel (fl. 1765–94), 258
Whitworth, Robert (1734–99), 167, 168
Williams, Sir Edward Leader (1828–1910), 279, 280
Williamson, James & Partners, 30, 34
Woodhouse, John (circa 1765–1830), 185
Yeoman, Thomas (circa 1708–81), 166

Architects

Bennet, T.P. & Co., 245
Casson, Sir Hugh (1910–), 146
Elmslie, E.W. (1819–89), 178
Gwynn, John (1713–86), 248
Hardwick, Philip (1792–1870), 193
Harris, Daniel (fl. circa 1793, d. 1835), 172
Harrison, Thomas (1744–1829), 265
Hollis, Charles (fl. 1820–30), 110
Hopper, Thomas (1776–1856), 28
Jones, Inigo (1573–1652), 39
Pritchard, Thomas Farnolls (1723–77), 238
Strahan, John (fl. circa 1725, d. circa 1740), 142
Thompson, Francis (circa 1794–1871), 17, 269

Contractors

Allen, Ralph, 142
Armstrong, Sir William (1810–1900), 82
Armstrong Whitworth, 205
Associated British Bridge Builders, 115
Bailey, Crawshay (1789–1872), 77
Betts, Edward Ladd (1815–72), 16, 232, 269, 283
Boulton and Watt, 154
Brassey, Thomas (1805–70), 16, 49, 51, 63, 190, 209, 212, 213, 232, 287
Brymbo Company, 61, 62
Carline, John (1730–93), 249
Carline, John (Junior) (1761–1835), 249
Cementation Co., 34
Cleveland Bridge & Engineering Co., 24, 108
Coalbrookdale Company, 224, 232, 240
Cochrane, John (1823–91), 18, 287
Copperhouse Foundry, Hayle, 120
Costain, Richard, 78
Costain Tarmac Joint Venture, 15
Cowlin, William & Sons, 124
Davidson, Matthew (circa 1770–1819), 48, 249
Davies, David (1818–90), 59, 60, 103
Dixon, John (1835–91), 26
Dorman Long, 285
Fairclough Civil Engineering, 204
Fairclough, Leonard, 285
Finch of Chepstow, 82
Hamilton Ironworks, Liverpool, 117
Hammond and Murray, 244
Harvey of Hayle, 138, 154
Hazledine, William (1763–1840), 11, 12, 48, 111, 177, 208, 262
Holme, Arthur Hill (1841–1912), 31
Hoof, William (1788–1855), 272, 276
Horseley Company, 198, 199
Howard, John, 115
Isca Ironworks, 118
Jackson, Thomas (1808–85), 16
Jones, Gethin (fl. 1870s), 32
King, C.W. (1830–1916), 31
Llanidloes Railway Foundry, 65
Mackenzie, William (1794–1851), 16, 49, 51, 177, 200, 213
McAlpine, Sir Alfred & Son, 35
McIlquham, James, 152
McIntosh, David (1799–1856), 209, 283
McIntosh, Hugh (1768–1840), 164
Nuttall, Edmund, Ltd., 51
Onions, John (1745–1819), 236
Onions, John (Junior) (circa 1779–1859), 233
Peto, Sir Samuel Morton (1809–89), 232
Pritchard, Daniel (circa 1777–1843), 272, 276
Provis, William (1792–1870), 206
Savin, Thomas (1826–89), 59, 90
Simpson, John (1755–1815), 48, 231, 250
Stephenson, John (1794–1848), 16, 49, 51, 209, 213
Stothert of Bath, 149

Contractors (continued)
Strachan, John (1848–1909), 52
Taylor Woodrow Construction, 245
Thorne, Alfred (b. 1867?), 27, 285
Tilley, John (d. circa 1796), 249
Townshend, Thomas (circa 1771–1846), 200, 209
Toye, John & Company, 221
Trubshaw, James (1777–1853), 209, 223, 266, 275
Walker, Thomas Andrew (1828–89), 103, 137
Ward, John (d. 1866), 90
Wharton & Warden, 22
Widnes Foundry Co., 27
Williams, John, 35, 61
Wilson, John (1772–1831), 11, 48, 206
Wimpey, George & Co., 66

Subject Index

Abercamlais Suspension Bridge, 77
Aberchalder Bridge, 143
Abermule Bridge, 60, 61, 62
Aberystwyth Cliff Railway, 56
Acton Bridge, 279
Acton Grange Viaduct, 282, 283–284
Albert Edward Bridge, nr Buildwas, 240
Almondsbury Tunnel, 136
Anderton Lift, 280–281
Aqueducts and Pipelines
- Avoncliff Aqueduct, 147, 153
- Barton Swing Aqueduct, 283
- Bewdley Pipeline Bridge, 229, 230
- Brynich Aqueduct, 76
- Chirk Aqueduct, 48–49
- Congleton Aqueduct, 260
- Dundas Aqueduct, 147, 152–153
- Elan Valley Aqueduct, 229, 230
- Engine Arm Aqueduct, 197–198, 201
- Hazlehurst Aqueduct, 272–273
- Longdon on Tern Aqueduct, 47, 92, 252–253
- Milford Aqueduct, 218
- Nantwich Aqueduct, 260
- Pontcysyllte Aqueduct, 45, 47–48, 51, 92, 258, 259
- Pont y Cafnau, 91–92
- Steward Aqueduct, 201
- Stretton Aqueduct, 207–208, 260
- Trent Aqueduct, Great Haywood, 217–218
- Twrch Aqueduct, 80–81
- Vyrnwy Aqueduct, 43–44

Ashford Carbonel Bridge, 228–229
Aust Overhead Cable Crossing, 115
Avon and Gloucestershire Railway, 171
Avoncliff Aqueduct, 147, 153
Bage's Mill, Shrewsbury, 244–245
Barmouth Viaduct, 37
Barnton Tunnel, 269
Barry Docks, 60, 103–104
Barton Swing Aqueduct, 283
Bath, 141, 142, 143, 144, 145, 149
Beaminster Road Tunnel, 209
Beeston Iron Lock, 259–260
Belvidere Bridge, Shrewsbury, 243–244
Berw Bridge, Pontypridd, 88–89
Bewdley Bridge, 231
Bewdley Pipeline Bridge, 229, 230
Bigsweir Bridge, 110
Binn Wall, 136–137
Birmingham and Derby Junction Railway, 183
Birmingham and environs, 191, 192, 194, 196, 197, 198, 200, 201, 202, 203, 205
Birmingham and Fazeley Canal, 195
Birmingham and Gloucester Railway, 170, 182–184
Birmingham and Liverpool Junction Canal, 206–207, 208, 215, 250, 254, 276
Birmingham Canal, 177, 194–196, 199, 200, 203, 205, 206, 215
Birnbeck Pier, Weston-super-Mare, 118–119
Blackland Mill Iron Roof, 157
Blaenau Ffestiniog Railway Tunnel, 32–33
Bosley Locks, 277
Box Tunnel, 130, 132–133
Bratch Locks, 216–217
Brecknock and Abergavenny Canal, 76
Bredwardine Bridge, 174–175
Bridges, Railway
- Acton Grange Viaduct, 282, 283–284
- Albert Edward Bridge, 240
- Barmouth Viaduct, 37
- Belvidere Bridge, 243–244
- Britannia Bridge, 17, 22–25

Bridges, Railway (continued)
Carmarthen Railway Bridge, 81
Cefn Coed y Cymmer Viaduct, 90
Cefn Viaduct, 51
Chepstow Railway Bridge, 24, 94, 171
Chippenham Viaduct, 130
Chirk Railway Viaduct, 49
Chorley Flying Arches, 82
Cnwclas Viaduct, 62–63
Conwy Tubular Bridge, 17, 22
Crumlin Viaduct, 70, 95, 105–107
Cwmbach Bridge, 93–94
Cynghordy Viaduct, 67
Dutton Viaduct, 282, 283
Frampton Mansell Viaduct, 170
Frodsham Viaduct, 279
Gothic Arch, Temple Meads, 131
Hawarden Swing Bridge, 18
Hengoed Viaduct, 95
Landore Viaduct, 81
Llansamlet Flying Arches, 82
Lledr Viaduct, 32
Llwchwr Viaduct, 81
Malltreath Viaduct, 17
Neath River Swing Bridge, 82
Newark Dyke Bridge, 106
Ogwen Viaduct, 17
Penkridge Viaduct, 190, 211, 212
Penmaenmawr Sea Viaduct, 20, 21–22
Penydarren Tramroad Bridges, 85
Pontypridd Viaduct, 86
Quaker's Yard Viaduct, 85, 86
Robertstown Tramway Bridge, 94–95
Royal Albert Bridge, Saltash, 24, 120, 171
Runcorn Railway Bridge, 286–287
Severn Railway Bridge, 165, 171
Tan-y-Bwlch Iron Bridge, 33
Victoria Bridge, Arley, 232, 240
Wicker Arches, Sheffield, 172
Winterbourne Viaducts, 139

Bridges, Road and Pedestrian
Abercamlais Suspension Bridge, 77
Aberchalder Bridge, 143
Abermule Bridge, 60, 61, 62
Acton Bridge, 279
Ashford Carbonel Bridge, 228–229
Berw Bridge, Pontypridd, 88–89
Bewdley Bridge, 231
Bigsweir Bridge, 110
Bredwardine Bridge, 174–175
Caerhowel Bridge, 60, 62
Cantlop Bridge, 242, 243
Castle Walk Footbridge, Shrewsbury, 245
Chepstow Bridge, 111
Chester Old Dee Bridge, 263, 265
Chetwynd Bridge, 224–225
Clifton Suspension Bridge, 119–121, 142
Coalport Bridge, 236–237
Conwy Arch Bridge, 14
Conwy Suspension Bridge, 13
Cound Arbour Bridge, 241
Crickhowell Bridge, 75
Cumberland Basin Swing Bridge, Bristol, 123
Cysyllte Ancient Bridge, 48
Devil's Bridge, 57–58
Dolauhirion Bridge, 67
Doldowlod Bridge, 65
Dutton Horse Bridge, 279, 283
Eaton Hall Iron Bridge, 261–262
English Bridge, Shrewsbury, 247–248
Essex Bridge, Great Haywood, 221–222
Galton Bridge, 201–202
Glanrhyd Bridge, 102–103
Great Bedwyn Skew Bridge, 147–148
Great Haywood Canal Bridge, 218
Grosvenor Bridge, Chester, 265–266
Hazlehurst Iron Bridge, 273
Hockenhull Packhorse Bridge, 262–263
Holt Ancient Bridge, 261
Holt Fleet Bridge, 181–182, 202
Iron Bridge, 237, 238–240
Kings Weston Lane Bridge, Bath, 149
Ladies Bridge, Wilcot, 147
Llandeilo Bridge, 77–78
Llandinam Bridge, 60
Llangollen Ancient Bridge, 44
Llangynidr Bridge, 75–76
Llanrwst Bridge, 39–40

Bridges, Road and Pedestrian (continued)
Mavesyn Ridware Bridge, 223–224
Menai Bridge, 10, 11–12, 24
Middlesbrough Transporter Bridge, 108
Monnow Bridge, Monmouth, 72–73
Montford Bridge, 249–250
Mor Brook Bridge, 233
Mythe Bridge, 60, 177–178, 202
New Bridge, Bath, 141–142
New Inn Bridge, 101–102
Newport Transporter Bridge, 108–109
Over Bridge, 161–162
Pant y Goitre Bridge, 74–75
Pont Carrog, 40–41
Pont Cilan, 41
Pont Dyfrydwy, 41
Pont y Gwaith, 85, 89
Pontypridd Bridge, 67. 85, 87–88
Powick Bridges, Worcester, 180–181
Pulteney Bridge, Bath, 145
Quay Pit Bridge, 177
Rabone Lane Canal Bridges, 198–199
Rainhill Skew Arch, 172
Rhosferrig Suspension Bridge, 77
River Dee Viaduct, 51
Runcorn–Widnes Bridge, 285–286
Severn Road Bridge, 114–116
Shugborough Hall Bridges, 220–221
Stanford Bridge, 186–187
Stowell Park Bridge, 65, 143, 148
Sydney Gardens Bridges, Bath, 149–150
Tipton Lift Bridge, 205–206
Top Lock Bridge, Bath, 149
Town and Hayhurst Bridges, 279
Town Bridge, Lechlade, 172–173
Tyne Bridge, Newcastle, 285
Victoria Bridge, Bath, 65, 142–143
Warkworth Bridge, Northumberland, 73
Warrington Bridge, 284–285
Warrington Transporter Bridge, 108
Wash House Lock Bridge, Bath, 149
Waterloo Bridge, Betws-y-coed, 10, 12–13
Welsh Bridge, Shrewsbury, 247, 248–249
Whitney Toll Bridge, 173–174
Wolseley Bridge, 222–223
Wye Bridge, Hereford, 175–177

Bridgewater Canal, 203, 219, 269

Bridgnorth, Castle Hill Cliff Railway, 234–236

Bridgnorth, St Mary Magdalene Church, 233–234

Bristol, 119, 121, 123, 124, 125, 127, 134, 135

Bristol and Bath Turnpike Trusts, 136, 139–141

Bristol and Exeter Railway, 135

Bristol and Gloucester Railway, 170

Bristol and South Wales Union Railway, 136–137

Bristol City Docks, 121–122

Bristol Waterworks, 127–129

Britannia Bridge, 17, 22–25

Bruce Tunnel, Kennet and Avon Canal, 147

Brunel's Drag Boat, Bristol, 122

Brynich Aqueduct, 76

Buildings, Industrial
Bage's Mill, Shrewsbury, 244–245
Blackland Mill Iron Roof, 157
Clock Warehouse, Stourport, 185–186
Tobacco Warehouses, Bristol, 124–125

Buildings, Other
Bridgnorth, St Mary Magdalene Church, 233–234
Madeley, St Michael's Church, 234
Malinslee, St Leonard's Church, 234

Burry Port and Gwendraeth Valley Railway, 96–97

Caban Coch Dam, 63–64, 229

Caerhowel Bridge, 60, 62

Caldon Canal, 270, 272, 274

Cambrian Coast Railway Lines, 36–37

Cambrian Railway, 51, 52, 57

Canals
Anderton Lift, 280–281
Beeston Iron Lock, 259–260
Birmingham, 177, 194–196, 199, 200, 203, 205, 206, 215
Birmingham and Fazeley, 195

Canals (continued)
Birmingham and Liverpool Junction, 206–207, 208, 215, 250, 254, 276
Bosley Locks, 277
Bratch Locks, 216–217
Brecknock and Abergavenny, 76
Bridgewater, 203, 219, 269
Caldon, 270, 272, 274
Chester, 45, 206, 250, 258–259
Chester Northgate Locks, 259
Coventry, 196, 220
Devizes Locks, 147, 153–154
Donnington Wood, 252
Dudley, 195, 202, 203
Ellesmere, 45, 47, 48, 49, 250, 258–259, 260
Exeter, 164
Froghall Basin, 270, 274–276
Gas Street Basin, Birmingham, 196–197
Glamorganshire, 70, 85, 86, 104
Gloucester & Sharpness, 129, 159, 163, 164–165, 166
Grand Trunk, 219, 257, 274
Grub Street Cutting, 254–255
Hay Inclined Plane, Coalport, 237–238, 252
Kennet and Avon, 146–148, 149, 150, 152, 153, 154, 159, 171
Ketley, 252
Kings Norton Stop Lock, 191–192
Macclesfield, 260, 270, 276–277
Manchester Ship, 43, 279, 282, 283
Monmouthshire, 107
Montgomeryshire, 61, 250
Oxford, 220
Peak Forest, 270, 276
Rogerstone Locks, 107
Rushall, 195
Shelmore Embankment, 207, 208–209, 255
Shrewsbury, 206, 250, 252
Shropshire, 237, 252
Shropshire Union, 207, 250–252, 258
Shropshire Union, Llangollen Branch, 45, 47, 48, 49, 250, 258
Smethwick Cutting, 200–201
Somerset Coal, 152, 153
Staffordshire and Worcestershire, 159, 185, 190, 194, 206, 215, 216, 217, 218, 220
Stourport Canal Terminus, 185–186, 215
Stratford-upon-Avon, 191,196
Stroudwater, 146, 159, 165, 166, 167
Swansea, 80
Tame Valley, 195
Tardebigge Locks, 153, 184–185
Thames and Severn, 146, 159, 166, 167–168, 170
Trent and Mersey, 190, 206, 215, 217, 218–220, 258, 269–270, 274, 276, 279, 280
Uttoxeter, 275
Warwick and Birmingham, 196
Weaver Navigation, 279–280
Widcombe Locks, 147, 149
Wombridge, 252
Woodseaves Cutting, 254
Worcester and Birmingham, 153, 159, 184, 191, 196
Wyrley and Essington, 195

Cantlop Bridge, 242–243
Carew Tide Mill, 95–96
Carmarthen Railway Bridge, 81
Castle Walk Footbridge, Shrewsbury, 245
Cefn Coed y Cymmer Viaduct, 90
Cefn Viaduct, 51
Central Wales Railway, 62, 67
Charmouth Road Tunnel, 209
Cheddleton Mills, 273–274
Cheltenham and Great Western Union Railway, 169
Chelvey Pumping Station, 129
Chepstow Bridge, 111
Chepstow Railway Bridge, 24, 94, 171
Chester, 263, 265, 267, 269
Chester and Holyhead Railway, 8, 15–17, 19, 21, 22, 25, 27, 211, 269
Chester Canal, 45, 206, 250, 258–259
Chester Northgate Locks, 259
Chester Old Dee Bridge, 263, 265

Chester Water Tower, 267–268
Chester Weir, 263–264
Chetwynd Bridge, 224–225
Chippenham Viaduct, 130
Chirk Aqueduct, 48–49
Chirk Canal Tunnel, 49
Chirk Railway Viaduct, 49
Chorley Flying Arches, 82
Claverton Pumping Station, 147, 150–151, 154
Clevedon Pier, 116–117, 118
Cliff Railways
  Aberystwyth Cliff Railway, 56
  Bridgnorth, Castle Hill Cliff Railway, 234–236
Clifton Suspension Bridge, 119–121, 142
Clock Warehouse, Stourport, 185–186
Cnwclas Viaduct, 62–63
Coalport Bridge, 236–237
Cob, The, Porthmadog, 33, 35–36
Coleham Pumping Station, Shrewsbury, 246–247
Congleton Aqueduct, 260
Conwy Arch Bridge, 14
Conwy Immersed Tube Tunnel, 14–15
Conwy Suspension Bridge, 13
Conwy Tubular Bridge, 17–22
Cound Arbour Bridge, 241
Coventry Canal, 196, 220
Crewe Railway Works, 19, 282
Crickhowell Bridge, 75
Crofton Pumping Station, 147, 154–155
Crumlin Viaduct, 70, 95, 105–107
Cumberland Basin, Bristol, 122
Cumberland Basin Swing Bridge, Bristol, 123
Curzon Street Station, Birmingham, 182, 192–194, 209
Cwmbach Railway Bridge, 93–94
Cynghordy Viaduct, 67
Cysyllte Ancient Bridge, 48
Dams, Weirs and Reservoirs
  Caban Coch Dam, 63–64, 229
  Chester Weir, 263–264
  Horseshoe Falls, 45–46
  Llyn Brianne Dam, 66
  Monmouth Forge Weir, 74
  Nant Hir Reservoir, 79–80
  Netham Dam, Bristol, 122
  Penarth Weir, 61–62
  Pulteney Weir, Bath, 145–146
  Upper Neuadd Dam, 79
  Usk Dam, 78–79
  Vyrnwy Dam, 41–43
  Ynys y Fro Reservoir, 107–108
Devil's Bridge, 57–58
Devizes Locks, 147, 153–154
Dinorwig Pumped Storage Scheme, 5, 28–31
Docks and Harbours
  Barry Docks, 60, 103–104
  Bristol City Docks, 121–122
  Cumberland Basin, Bristol, 122
  Gloucester Docks, 163–164
  Holyhead Dry Dock, 7
  Holyhead Harbour, 7–9, 25
  Holyhead Inner Harbour, 7
  Holyhead New Harbour, 7
  Lydney Docks, 160, 171
  Sharpness Docks, 159, 164
  South Entrance Lock, Bristol, 122, 123
Dolauhirion Bridge, 67
Doldowlod Bridge, 65
Donnington Wood Canal, 252
Drainage
  Coleham Pumping Station, Shrewsbury, 246–247
  River Dee Channel, 268–269
Dudley Canal, 195, 202, 203
Dudley Canal Tunnel, 202, 203–204
Dundas Aqueduct, 147, 152–153
Dutton Horse Bridge, 279, 283
Dutton Viaduct, 282, 283
Eaton Hall Iron Bridge, 261–262
Elan Valley Aqueduct, 229, 230

Elan Valley Reservoirs, 55, 63
Ellesmere Canal, 45, 47, 48, 49, 250, 258–259, 260
Engine Arm Aqueduct, 197–198, 201
English Bridge, Shrewsbury, 247–248
Essex Bridge, Great Haywood, 221–222
Exeter Canal, 164
Ffestiniog Pumped Storage Scheme, 33, 34–35
Ffestiniog Railway, 28, 32, 33–34, 35, 36, 37, 38
Frampton Mansell Viaduct, 170
Frodsham Viaduct, 279
Froghall Basin, 270, 274–276
Galton Bridge, 201–202
Garth Pier, Bangor, 27
Gas Street Basin, Birmingham, 196–197
Glamorganshire Canal, 70, 85, 86, 104
Glanrhyd Bridge, 102–103
Gloucester and Sharpness Canal, 129, 159, 163, 164–165, 166
Gloucester Docks, 163–164
Gothic Arch, Temple Meads, 131
Grand Junction Railway, 190, 192, 209–211, 212, 257, 281–282, 283
Grand Trunk Canal, 219, 257, 274
Great Bedwyn Skew Bridge, 147–148
Great Haywood Canal Bridge, 218
Great Malvern Station, 178–180
Great Western Railway, 53, 57, 71, 81, 97, 113, 129–132, 134, 135, 136, 137, 138, 146, 150, 151, 160, 165, 194
Green Park Station, Bath, 143–144
Grosvenor Bridge, Chester, 265–266
Grub Street Cutting, 254–255
Harecastle Tunnels, 219, 257, 269, 271–272
Hawarden Swing Bridge, 18
Hay Inclined Plane, Coalport, 237–238, 252
Hazlehurst Aqueduct, 272–273
Hazlehurst Iron Bridge, 273
Hengoed Viaduct, 95
Hockenhull Packhorse Bridge, 262–263
Holt Ancient Bridge, 261
Holt Fleet Bridge, 181–182, 202
Holyhead Admiralty Pier, 7
Holyhead Breakwater, 7
Holyhead Dry Dock, 7
Holyhead Harbour, 7–9, 25
Holyhead Inner Harbour, 7
Holyhead New Harbour, 7
Holyhead Railway Station, 25–26
Holyhead Road, 5, 9–11, 15, 45, 189, 214–215, 247, 249
Holyhead South Pier, 7
Horseshoe Falls, 45–46
Huddersfield Railway Station, 25
Iron Bridge, 237, 238–240
Jodrell Bank Radio Telescope, 277–278
Kennet and Avon Canal, 146–148, 149, 150, 152, 153, 154, 159, 171
Ketley Canal, 252
Kings Norton Stop Lock, 191–192
King's Stanley Mill, 166–167
Kings Weston Lane Bridge, Bath, 149
Ladies Bridge, Kennet and Avon Canal, 147
Landore Viaduct, 81
Lickey Incline, 183
Lighthouses
- South Stack Lighthouse, 6–7
- Whiteford Point Lighthouse, 99–100, 104

Liverpool and Manchester Railway, 15, 192, 209, 282
Llanberis Water Wheel, 30
Llandeilo Bridge, 77–78
Llandinam Bridge, 60
Llandudno Pier, 26–27
Llanelli and Mynydd Mawr Railway, 97–98
Llanelli Railway, 98–99

Llangollen Ancient Bridge, 44
Llangynidr Bridge, 75–76
Llanrwst Bridge, 39–40
Llansamlet Flying Arches, 82
Llechwedd Slate Caverns, 35
Lledr Viaduct, 32
Llwchwr Viaduct, 81
Llyn Brianne Dam, 66
London and Birmingham Railway, 192
London and North Western Railway, 9, 19, 25, 32, 98, 211, 250, 286
Longdon on Tern Aqueduct, 47, 92, 252–253
Lydney Docks, 160, 171
Macclesfield Canal, 260, 270, 276–277
Madeley, St. Michael's Church, 234
Malinslee, St. Leonard's Church, 234
Manchester Airport, 287
Manchester and Birmingham Railway, 211
Manchester, Sheffield and Lincolnshire Railway, 18
Manchester Ship Canal, 43, 279, 282, 283
Marchlyn Mawr Reservoir, 29
Mavesyn Ridware Bridge, 223–224
Melingriffith Water Pump, 104–105
Menai Bridge, 10, 11–12, 24
Middlesbrough Transporter Bridge, 108
Midland Railway, 143, 160, 165, 170, 183
Milford Aqueduct, 218
Moelwyn Tunnel, 33
Monmouth Forge Generating Station, 73–74
Monmouth Forge Weir, 74
Monmouthshire Canal, 107
Monnow Bridge, Monmouth, 72–73
Montford Bridge, 249–250
Montgomeryshire Canal, 61, 250
Mor Brook Bridge, 233
Mythe Bridge, 60, 177–178, 202
Nant Ffrancon Pass, 10
Nant Hir Reservoir, 79–80
Nantwich Aqueduct, 260
Neath River Swing Bridge, 82
Netham Dam, Bristol, 122
Netherton Canal Tunnel, 202–203
Newark Dyke Bridge, 106
New Bridge, Bath, 141–142
New Inn Bridge, 101–102
Newport Abergavenny and Hereford Railway, 106
Newport Transporter Bridge, 108–109
Newtown and Machynlleth Railway, 58
North Wales Coastal Defences, 19–21
Ogwen Viaduct, 17
Oldbury Nuclear Power Station, 115
Over Bridge, 161–162
Oxford Canal, 220
Pant y Goitre Bridge, 74–75
Peak Forest Canal, 270, 276
Penarth Weir, 61–62
Penkridge Viaduct, 190, 211, 212
Penmaenbach Tunnel, 20
Penmaenmawr Sea Viaduct, 20, 21–22
Penmaenmawr Tunnel, 20
Penmaenrhos Tunnel, 20
Penrhyn Castle Industrial Museum, 28
Penrhyn Railway, 27–28
Penydarren Tramroad, 83–85, 86
Penydarren Tramroad Bridges, 85
Piers and Breakwaters
    Admiralty Pier, Holyhead, 7
    Birnbeck Pier, Weston-super-Mare, 118–119
    Clevedon Pier, 116–117, 118
    Garth Pier, Bangor, 27
    Holyhead Breakwater, 7
    Holyhead South Pier, 7
    Llandudno Pier, 26–27
    Victoria Pier, Colwyn Bay, 27
Pont Carrog, 40–41
Pont Cilan, 41

Pontcysyllte Aqueduct, 45, 47–48, 51, 92, 258, 259

Pont Dyfrydwy, 41

Pont y Cafnau, 91–92

Pont y Gwaith, 85, 89

Pontypridd Bridge, 67, 85, 87–88

Pontypridd Viaduct, 86

Power Generating Stations
- Dinorwig Pumped Storage, 5, 28–31
- Ffestiniog Pumped Storage, 33, 34–35
- Monmouth Forge Generating Station, 73–74
- Oldbury Nuclear Power Station, 115

Powick Bridges, Worcester, 180–181

Prestatyn Station, 19

Preston Brook Tunnel, 270

Pulteney Bridge, Bath, 145

Pulteney Weir, Bath, 145–146

Quakers Yard Viaduct, 85, 86

Quay Pit Bridge, 177

Rabone Lane Canal Bridges, 198–199

Railways
- Avon and Gloucestershire, 171
- Birmingham and Derby Junction, 183
- Birmingham and Gloucester, 170, 182–184
- Bristol and Exeter, 135
- Bristol and Gloucester, 170
- Bristol and South Wales Union, 136–137
- Burry Port and Gwendraeth Valley, 96–97
- Cambrian, 51, 52, 57
- Cambrian Coast Lines, 36–37
- Central Wales, 62, 67
- Cheltenham and Great Western Union, 169
- Chester and Holyhead, 8, 15–17, 19, 21, 22, 25, 27, 211, 269
- Crewe Works, 19, 282
- Ffestiniog, 28, 32, 33–34, 35, 36, 37, 38
- Grand Junction, 190, 192, 209–211, 212, 257, 281–282, 283
- Great Western, 53, 57, 71, 81, 97, 113, 129–132, 134, 135, 136, 137, 138, 146, 150, 151, 160, 165, 194
- Lickey Incline, 183
- Liverpool and Manchester, 15, 192, 209, 282
- Llanelli, 98–99
- Llanelli and Mynydd Mawr, 97–98
- London and Birmingham, 192
- London and North Western, 9, 19, 25, 32, 98, 211, 250, 286
- Manchester and Birmingham, 211
- Manchester, Sheffield and Lincolnshire, 18
- Midland, 143, 160, 165, 170, 183
- Newport Abergavenny and Hereford, 106
- Newtown and Machynlleth, 58
- Penrhyn, 27–28
- Penydarren Tramroad, 83–85, 86
- Ralph Allen's Waggonway, 147
- Shrewsbury and Chester, 49, 51
- Snowdon Mountain, 31–32
- Somerset and Dorset Joint, 143
- South Wales, 81–83, 97, 116, 171
- South Wales Direct, 138–139
- Stockton and Darlington, 132, 192
- Surrey Iron, 97
- Swansea and Mumbles, 100–101
- Swindon to Severn Tunnel Junction, 169–172
- Taff Vale, 82, 85–87
- Talyllyn, 37–39
- Trent Valley Railway, 190, 211, 212
- Welshpool and Llanfair, 51–53
- Vale of Rheidol, 37, 57
- Vale of Towy, 98

Railway Stations
- Curzon Street, Birmingham, 182, 192–194, 209
- Great Malvern, 178–180
- Green Park, Bath, 143–144
- Holyhead, 25–26
- Huddersfield, 25
- Prestatyn, 19
- Temple Meads (Old), Bristol, 134–135
- Temple Meads (Present), Bristol, 135
- York, 135

Rainhill Skew Arch, 172

Ralph Allen's Waggonway, 147

Rhosferrig Suspension Bridge, 77

Rhuddlan Castle, 21

River Dee Channel, 268–269

River Dee Viaduct, 51

Roads

Bristol and Bath Turnpike Trusts, 136, 139–141

Cob, The, Porthmadog, 33, 35–36

Holyhead Road, 5, 9–11, 15, 45, 189, 214–215, 247, 249

Nant Ffrancon Pass, 10

Roman Road, Blackpool Bridge, 160–161

Stanley Embankment, 10, 17

Robertstown Tramway Bridge, 94–95

Rogerstone Locks, 107

Roman Road, Blackpool Bridge, 160–161

Roman Water Supply and Drainage, Bath, 144–145

Royal Albert Bridge, Saltash, 24, 120, 171

Runcorn Railway Bridge, 286–287

Runcorn–Widnes Bridge, 285–286

Rushall Canal, 195

Saltersford Tunnel, 270

Sapperton Canal Tunnel, 168–169, 170

Sapperton Railway Tunnels, 170

Sea Defences

Binn Wall, 136–137

Cob, The, Porthmadog, 33, 35–36

North Wales Coastal Defences, 19–21

Penmaenmawr Sea Viaduct, 20, 21–22

River Dee Channel, 268–269

Severn Railway Bridge, 165, 171

Severn Road Bridge, 114–116

Severn Tunnel, 82, 104, 136, 137–138, 139, 171

Sharpness Docks, 159, 164

Shelmore Embankment, 207, 208–209, 255

Shrewsbury and Chester Railway, 49, 51

Shrewsbury Canal, 206, 250, 252

Shropshire Canal, 237, 252

Shropshire Union Canal, 207, 250–252, 258

Shropshire Union Canal, Llangollen Branch, 45, 47, 48, 49, 250, 258

Shugborough Hall Bridges, 220–221

Shugborough Tunnel, 212–213

Silbury Hill, 156–157

Smethwick Cutting, 200–201

Smethwick Engine, Birmingham, 154

Snowdon Mountain Railway, 31–32

Sodbury Tunnel, 139

Somerset and Dorset Joint Railway, 143

Somerset Coal Canal, 152, 153

South Entrance Lock, Bristol, 122, 123

South Stack Lighthouse, 6–7

South Wales Direct Railway, 138–139

South Wales Railway, 81–83, 97, 116, 171

SS *Great Britain*, 125–127

Staffordshire and Worcestershire Canal, 159, 185, 190, 194, 206, 215, 216, 217, 218, 220

Stanford Bridge, 186–187

Stanley Embankment, 10, 17

Steward Aqueduct, 201

Stockton and Darlington Railway, 132, 192

Stourport Canal Terminus, 185–186, 215

Stowell Park Bridge, 65, 143, 148

Stratford upon Avon Canal, 191, 196

Stretton Aqueduct, 207–208, 260

Stroudwater Canal, 146, 159, 165, 166, 167

Surrey Iron Railway, 97

Swansea and Mumbles Railway, 100–101

Swansea Canal, 80

Swindon Railway Works, 130

Swindon to Severn Tunnel Junction Railway, 169–172

Sydney Gardens Bridges, Bath, 149–150

Taff Vale Railway, 82, 85–87

Talerddig Cutting, 58–59

Talyllyn Railway, 37–39

Tame Valley Canal, 195

Tan-y-Bwlch Iron Bridge, 33
Tardebigge Locks, 153, 184–185
Temple Meads (Old) Station, 134–135
Temple Meads (Present) Station, 135
Thames and Severn Canal, 146, 159, 166, 167–168, 170
Three Railways near Llanelli, 96–99
Tipton Lift Bridge, 205–206
Tobacco Warehouses, Bristol, 124–125
Tollhouses, 10
Top Lock Bridge, Bath, 149
Town and Hayhurst Bridges, 279
Town Bridge, Lechlade, 172–173
Trent Aqueduct, Great Haywood, 217–218
Trent and Mersey Canal, 190, 206, 215, 217, 218–220, 258, 269–270, 274, 276, 279, 280
Trent Valley Railway, 190, 211, 212
Tunnels, Canal
  Barnton, 269
  Bruce, 147
  Chirk, 49
  Dudley, 202, 203–204
  Harecastle, 219, 257, 269, 271–272
  Netherton, 202–203
  Preston Brook, 270
  Saltersford, 270
  Sapperton, 168–169, 170
Tunnels, Railway
  Almondsbury, 136
  Blaenau Ffestiniog, 31–32
  Box, 130, 132–133
  Moelwyn, 33
  Penmaenbach, 20
  Penmaenmawr, 20
  Penmaenrhos, 20
  Sapperton, 170
  Severn, 82, 104, 136, 137–138, 139, 171
  Shugborough, 212–213
  Sodbury, 139
Tunnels, Road
  Beaminster, 209
  Charmouth, 209
  Conwy Immersed Tube, 14–15
Twrch Aqueduct, 80–81
Tyne Bridge, Newcastle, 285
Upper Neuadd Dam, 79
Usk Dam, 78–79
Uttoxeter Canal, 275
Vale of Rheidol Railway, 37, 57
Vale of Towy Railway, 98
Victoria Bridge, Arley, 232, 240
Victoria Bridge, Bath, 65, 142–143
Victoria Pier, Colwyn Bay, 27
Vyrnwy Aqueduct, 43–44
Vyrnwy Dam, 41–43
Warkworth Bridge, Northumberland, 73
Warrington Bridge, 284–285
Warrington Transporter Bridge, 108
Warwick and Birmingham Canal, 196
Wash House Lock Bridge, Bath, 149
Water supply
  Bath Roman Water Supply and Drainage, 144–145
  Bristol Waterworks, 127–129
  Chelvey Pumping Station, Bristol, 129
  Chester Water Tower, 267–268
  Claverton Pumping Station, Kennet and Avon Canal, 147, 150–151, 154
  Crofton Pumping Station, Kennet and Avon Canal, 147, 154–155
  Melingriffith Water Pump, 104–105
  Vyrnwy, 41–44
Waterloo Bridge, Betws-y-coed, 10, 12–13
Water-mills
  Carew Tide Mill, 95–96
  Cheddleton Mills, 273–274
  King's Stanley Mill, 166–167
  Woodbridge Tide Mill, Suffolk, 96
Weaver Navigation, 279–280
Welsh Bridge, Shrewsbury, 247, 248–249
Welshpool and Llanfair Light Railway, 51–53
Whiteford Point Lighthouse, 99–100, 104
Whitney Toll Bridge, 173–174

Wicker Arches, Sheffield, 172
Widcombe Locks, Bath, 147, 149
Winterbourne Viaducts, 139
Wolseley Bridge, 222–223
Wombridge Canal, 252
Woodbridge Tide Mill, Suffolk, 96
Woodseaves Cutting, 254
Worcester and Birmingham Canal, 153, 159, 184, 191, 196
Wye Bridge, Hereford, 175–177
Wyrley and Essington Canal, 195
Ynys y Fro Reservoir, 107–108
York Station, 135